HIGH-BRIGHTNESS BEAMS FOR ADVANCED ACCELERATOR APPLICATIONS

CONFERENCE PROCEEDINGS NO. 253

PARTICLES AND FIELDS SERIES 47

HIGH-BRIGHTNESS BEAMS FOR ADVANCED ACCELERATOR APPLICATIONS

COLLEGE PARK, MD 1991

EDITORS:
WILLIAM W. DESTLER
& SAMAR K. GUHARAY
UNIVERSITY OF MARYLAND

American Institute of Physics   New York

Authorization to photocopy items for internal or personal use, beyond the free copying permitted under the 1978 U.S. Copyright Law (see statement below), is granted by the American Institute of Physics for users registered with the Copyright Clearance Center (CCC) Transactional Reporting Service, provided that the base fee of $2.00 per copy is paid directly to CCC, 27 Congress St., Salem, MA 01970. For those organizations that have been granted a photocopy license by CCC, a separate system of payment has been arranged. The fee code for users of the Transactional Reporting Service is: 0094-243X/87 $2.00.

© 1992 American Institute of Physics.

Individual readers of this volume and nonprofit libraries, acting for them, are permitted to make fair use of the material in it, such as copying an article for use in teaching or research. Permission is granted to quote from this volume in scientific work with the customary acknowledgment of the source. To reprint a figure, table, or other excerpt requires the consent of one of the original authors and notification to AIP. Republication or systematic or multiple reproduction of any material in this volume is permitted only under license from AIP. Address inquiries to Series Editor, AIP Conference Proceedings, AIP, 335 East 45th Street, New York, NY 10017-3483.

L.C. Catalog Card No. 92-52705
ISBN 0-88318-947-X
DOE CONF-9106260

Printed in the United States of America.

CONTENTS

Preface vii

The Emittance Concept 1
  J. D. Lawson
Early Studies on Intensity Limitations in Proton Linear Accelerators 11
  P. Lapostolle
Emittance Growth from Space-Charge Forces 21
  T. P. Wangler
Advances in the Theory of Charged Particle Beam Transport 41
  A. J. Dragt
Solitons and Particle Beams 42
  J. J. Bisognano
Experimental Studies of Emittance Growth in a Nonuniform, Mismatched, and
Misaligned Space-Charge Dominated Beam in a Solenoid Channel 47
  D. Kehne, M. Reiser, and H. Rudd
Studies of Longitudinal Beam Compression and Resistive-Wall Instability 57
  J. G. Wang, M. Reiser, D. X. Wang, and W. M. Guo
Studies of Bright Beam Transport by the LBL MFE Group 66
  O. A. Anderson, L. Soroka, and J. W. Kwan
High-Brightness H^- Beam Transport Using ESQ Lenses 67
  S. K. Guharay, C. K. Allen, and M. Reiser
The Influence of the Beam Plasma on the Emittance of Intense and Space-Charge
Compensated Beams 77
  T. Weiss
Ultra-Intense Laser Interactions with Beams and Plasmas 87
  P. Sprangle and E. Esarey
On Possibilities of Fast Cooling of Heavy Particle Beams 103
  Y. S. Derbenev
The Performance of the Tevatron Collider at Fermilab 111
  N. M. Gelfand
High Brightness and the SSC 122
  D. A. Edwards and M. J. Syphers
Intense Beams at the Micron Level for the Next Linear Collider 129
  J. T. Seeman
Advanced High-Brightness Ion RF Accelerator Applications in the Nuclear
Energy Arena 139
  R. A. Jameson
Non-Liouvillean Method Applied to Heavy Ion Fusion 149
  I. Hofmann
MBE-4 Experiments with Bright Cesium$^+$ Beams 160
  T. J. Fessenden
Beam Quality and Emittance in Free-Electron Lasers 170
  C. W. Roberson and B. Hafizi

The Los Alamos High-Brightness Photoinjector 182
    P. G. O'Shea
H⁻ Ion Sources 193
    J. G. Alessi
Saturnus: The UCLA Compact High-Brightness Linac 206
    C. Pellegrini
High Power Microwave Sources for Advanced Accelerators 213
    V. L. Granatstein, P. E. Latham, W. Lawson, W. Main, M. Reiser, C. D.
    Striffler, and S. Tantawi
Panel Discussion 227
    M. Reiser
Author Index 237

Preface

The generation and transport of high-brightness charged particle beams has become increasingly important as beam requirements for advanced accelerators and free-electron lasers have become more and more stringent. As a result, research in this area has grown steadily over the past two decades. The Symposium on High-Brightness Beams for Advanced Accelerator Applications, the first scientific meeting dedicated to this area of research, was held at the University of Maryland on June 6–7, 1991. The Symposium, hosted by the University of Maryland with organizational and financial support from the Department of Energy and the University of Maryland, brought together 92 researchers from all major accelerator laboratories in the United States and from a number of laboratories from Europe and Asia. Representatives from abroad included participants from TRIUMF in Canada, GSI in Germany, Rutherford-Appleton Laboratory in the United Kingdom, Niigata University and Tokyo Institute of Technology in Japan, and USTEK in China. In addition to the obvious objective of bringing together researchers in the area of high-brightness beams for accelerator applications, the Symposium was also organized in honor of Professor Martin Reiser's 60th birthday, and a banquet celebrating Professor Reiser's many years of distinguished contributions to the field was held on the evening of the first day of the Symposium. A number of photographs taken at the banquet are included on the pages immediately following this preface.

The Symposium included 5 sessions with a total of 21 oral presentations and a panel discussion. The papers included in these Proceedings cover topics ranging from fundamental emittance theory to theoretical and experimental studies of beam transport in advanced accelerator systems. The application of high-brightness beams in advanced accelerators for high-energy physics, heavy ion fusion, free-electron lasers, and radioactive waste transmutation was also a principal focus of the Symposium, and a number of papers in this volume are related to these applications.

The initial talks at the Symposium provided important historical perspectives from two leaders in the field of charged particle beam physics and accelerators. In his paper, J. D. Lawson offered a stimulating discussion of the concept of emittance. This paper, which he has announced is his last professional contribution, caps a distinguished career in which he has made seminal contributions to a number of areas, including particle accelerators and plasma physics. Dr. Lawson's paper was followed by a discussion of the pioneering early studies on the limitations on beam intensity in proton linear accelerators by Pierre Lapostolle. These studies were among the first to include space-charge effects in codes used to describe beam dynamics in accelerator systems.

A number of additional papers covered theoretical and experimental research on emittance growth and beam transport in accelerator systems. Thomas Wangler reported studies at the Los Alamos National Laboratory of emittance growth from space-charge forces in both transport and acceleration channels, and Alex Dragt of the University of Maryland reported on advanced theories for describing charged particle beam transport including the calculation and correction of aberrations resulting from nonlinear forces. In addition, Joseph Bisognano of CEBAF discussed soliton-like wave propagation in high-brightness beams. Kehne *et al.* of the University of Maryland reported studies of emittance growth due to beam mismatch, and Wang *et al.* also of the University of Maryland discussed studies of longitudinal beam compression and the resistive wall instability. Anderson *et al.* of LBL reported on new analytical and theoretical tools for designing practical electrostatic quadrupole (ESQ) transport systems, and Guharay *et al.* of the University of Maryland reported on particle simulation results and a novel design for an ESQ lens system to focus a high-brightness H^- beam. Thomas Weis of the University of Frankfurt, Germany, who was unable to attend the Symposium, contributed a paper on space-charge neutralized beams. Sprangle *et al.* of NRL discussed the interaction of intense laser pulses with beams and plasmas with possible applications in a number of areas including the radiative cooling of electron beams.

Finally, a paper on other topics relating to beam cooling had been contributed by Derbenev from the University of Michigan.

A number of contributors report on the design and construction of high-brightness beam sources and transport systems associated with advanced accelerators for high-energy physics experiments. Norman Gelfand, e.g., reported on the performance of the Tevatron Collider at Fermilab, including a discussion of attempts to increase the luminosity of the collider. Edwards and Syphers of the SSC discussed plans for emittance preservation in the design of the SSC, and John Seeman of SLAC discussed requirements for high-brightness beams in the next linear collider.

Beam generation, acceleration, and transport requirements in advanced accelerators designed for other applications were the subject of a number of additional papers in these Proceedings. Robert Jameson from LANL, e.g., reported on beam requirements for a next-generation proton accelerator designed for the accelerator-driven transmutation of radioactive waste. Ingo Hofmann of GSI discussed the increase of phase space density in a heavy ion fusion driver by means of a non-Liouvillean technique using photoionization of singly charged heavy ions. Tom Fessenden of LBL reported on experiments with high-brightness cesium beams intended to model much of the accelerator physics in the electrostatically focused section of a heavy ion fusion driver. Roberson et al. of ONR discusses the scaling of beam requirements for free-electron lasers designed for operation at short wavelengths and low voltages. Patrick O'Shea of LANL reported on a high-brightness photoinjector system developed for the Los Alamos free-electron laser project. James Alessi of Brookhaven reported on H^- sources and attempts to model extraction optics. Finally, Claudio Pellegrini discussed the design and expected beam characteristics of a compact high-brightness linac under construction at UCLA for studies of free-electron lasers and beam-plasma interactions. In a related paper, Victor Granatstein et al. discussed high power microwave sources for linear colliders and other applications.

In addition to the papers included in the Proceedings, a report on the panel discussion that concluded the Symposium has been contributed by Professor Reiser. In this report, Professor Reiser presented a digest of the presentations at the Symposium by comparing future advanced accelerator beam requirements (SSC, NLC, HIF, etc.) with the present state of the art, and highlights the beam physics research issues addressed at the Symposium. Also included in this panel report are brief comments by Robert Gluckstern and Pierre Lapostolle and a review of the history of non-Liouvillean injection by Ronald Martin. A full list of participants is included in the Appendix.

It is clear from the reports included in these Proceedings that interest in the generation, transport, and acceleration of high-brightness beams is rapidly increasing and that the Symposium served a very useful purpose in bringing together researchers in this area from around the world. It is hoped, therefore, that this first Symposium on High-Brightness Beams for Advanced Accelerator Applications will be followed in the not-too-distant future by a second symposium at which progress over the next few years can be reported. Special thanks are due to Dr. Terry Godlove, Dr. David Sutter, Dr. Charles Roberson, Dr. Robert Gluckstern, and Dr. Martin Reiser for chairing the sessions and keeping the compact, eventful meeting on schedule. We are pleased to acknowledge support from the U. S. Department of Energy, Office of High-Energy Physics, and the University of Maryland that was instrumental in bringing about the Symposium.

William W. Destler
Samar K. Guharay
Editors

THE EMITTANCE CONCEPT

J D Lawson
Rutherford Appleton Laboratory, Chilton, Oxon, OX11 0QX, UK

ABSTRACT

An informal descriptive account is first given of the emittance concept and its underlying physical basis. This is followed by a discussion of the connection between emittance and entropy, and a number of questions relating to problems of current interest concerning such topics as emittance growth and equipartition between different degrees of freedom are raised. Although no new results are obtained, it is hoped that the discussion may be helpful in the search for new insights. The paper differs from that presented at the conference, and contains ideas which arose in discussion with T P Wangler at Los Alamos after the conference.

INTRODUCTION

The term 'emittance' seems to have emerged in the early fifties, soon after the advent of alternating gradient focusing, and the general use of matrix techniques in accelerator design. It was preceded by the concept of 'admittance', defined in 1952 by Sigurgeirsson[1] for an alternating gradient synchrotron as

$$A = \iint \Omega(x, z) dx dz \qquad (1)$$

"where Ω is the solid angle, within which the direction of motion for a particle has to fall if it is to remain in the synchrotron without striking the walls of the vacuum chamber." In his discussion, which is in the context of the synchrotron, z is the vertical direction and x is the radial distance from the equilibrium orbit, neglecting the small curvature. (Here we use y and x for these two quantities). He shows that for symmetry about the xy and zy planes (his notation) the x and z variables can be separated, and that the bounding contour in the two directions is elliptical, and further that the area of the ellipses does not vary along the length of the beam. In general the areas of the ellipses corresponding to the two planes are different. The extension to emittance, as a figure of merit defining a beam with particles having coordinates that would just fill such an ellipse is straightforward. Further extension of the concept to the longitudinal direction is not difficult as discussed below.

The emittance was soon seen to be a convenient figure of merit for ion sources, though it was immediately recognized that a single number cannot contain information about the distribution of points projected on the xx' or yy' planes. This difficulty is particularly evident when aberrations are present, so that the distribution is not elliptical. For distributions that are roughly elliptical, conventions such as defining the emittance as the area of the smallest ellipse containing say 95% of the points are convenient, if not elegant. In modern usage the emittance is the area of the appropriate area divided by π.

Before discussing further some of the complications associated with the idea of emittance, some general observations about the properties of ensembles of particles in a Hamiltonian system without collisions will be summarized.

LIOUVILLE'S THEOREM

This theorem applies to ensembles of particles moving in a conservative system, under the action of an external potential, plus that arising from the smoothed-out self-fields of the particles themselves. By 'smoothed-out' is meant that individual particle-particle collisions are excluded; only the average fields, which do not depend on the positions of the individual particles, are included. The theorem, which is simply proved from Hamilton's equations and continuity, states that the density in 6 dimensional phase-space in the neighbourhood of any chosen point remains constant. From this it follows that if a surface is drawn around a portion of the phase-space 'fluid', its volume remains invariant though its shape can vary.

It is this invariance that makes the emittance a useful concept, though it is important to distinguish properties associated with the overall distribution from those associated with its projections. First we note that the variables xx' and yy' associated with the transverse emittance are not canonical, so that xx' space is not strictly 'phase-space'. In paraxial approximation, however, $x' = p_x/p_z$, so that $\beta\gamma x'$ and x are canonical. It is readily shown that for a <u>linear</u> focusing system, in which x, y and z motion are <u>decoupled</u>, projected areas enclosing a fixed number of points (each corresponding to a particle) are conserved. Furthermore, shapes are conserved apart from rotation and stretching. In particular, ellipses remain elliptical. These features are <u>not</u> present in linear systems with coupling between x and y directions, (for example, skew quadrupoles, or quadrupole + solenoid). For linear systems with axial symmetry, such as solenoids or conventional magnetic lenses, the x and y motion can be decoupled by making a transformation to the Larmor frame, which rotates with angular velocity $\omega_L = -eB_z/2\gamma m_0$.

For monoenergetic beams, with uniform line density in the direction of propagation, the transverse emittance can be determined from a knowledge of the particle positions and velocities in a small slice of beam of length Δz. If, however, there is structure in the z-direction, as is the case in a linear accelerator or synchrotron, the particles in a complete bunch need to be considered. The length of the bunch is just one wavelength in a linac, or the circumference C divided by the operating harmonic number n in a synchrotron. Momenta can be measured with respect to a particle moving at a velocity $f\lambda$ or fC/n. Alternatively, particles can be specified in terms of their phase with respect to the accelerating field and its derivative, ϕ and $\dot{\phi}$. In such cyclic systems any particles leaving the front of the bunch are replaced by others entering at the rear.

In this discussion some elementary properties of emittance for linear systems have been summarized; before proceeding it is necessary to attempt a definition.

DEFINITIONS OF EMITTANCE

The progress of a beam can be represented in general by the evolution of a distribution of points in six-dimensional phase space. In most beams the system is conservative, (though a notable exception is an electron beam or very high energy proton beam where synchrotron radiation is emitted). The question now is how best to quantify this distribution in terms of a single number. A heuristic method is to take a value proportional to the minimum sized hyper-ellipsoid (or ellipse for projections involving one spatial co-ordinate) circumscribing a certain fraction of the points. More precise is a description in terms of moments. The r.m.s. emittance $\bar{\varepsilon}$ of a projection in the xx' plane may be defined as [2,3]

$$\bar{\varepsilon}^2 = [16]\left(<x^2><x'^2> - <xx'>^2\right) \qquad (2)$$

The factor 16 is introduced in ref. 2 but not in ref. 3. Its purpose is to ensure parity with the earlier alternative definition in which the area of uniformly populated ellipse in xx' space is taken as $\pi\varepsilon$. (See for example ref. 4).

The following points are well established:

1) For a linear system in which x, y and z motion is decoupled, the normalized emittance $\beta\gamma\bar{\varepsilon}$ is invariant. In the presence of space-charge there are a limited number of distributions which produce a uniform projected distribution with sharp edge, and hence provide a linear contribution to the focusing; the best known of these is that of Kapchinskij and Vladimirskij, (K-V), discussed further below. The envelope equation for such beams, known as the K-V equation, is, for a drifting unaccelerated beam,

4 The Emittance Concept

$$\ddot{x} + k(x)x - \frac{\bar{\varepsilon}_x^2}{x^3} - \frac{2K}{x+y} = 0 \qquad (3)$$

where k(x) represents the focusing force and K is the dimensionless perveance.

2) For beams with non-uniform charge density, so that the contribution to focusing is non-linear, the K-V envelope equations are still valid provided that a) x and y are interpreted as r.m.s. values, b) the distribution has elliptical symmetry[3].

3) The value of $\bar{\varepsilon}$ is not, however, invariant. Although the value $d\bar{\varepsilon}/dz$ for a given situation can be found, the way that $\bar{\varepsilon}$ varies cannot be determined.

Most of the discussion so far has been confined to linear systems with separable variables. When the variables are coupled there are further invariants, corresponding to Eq. 2 though these are considerably more complicated in form[5].

So far, precise definition has been given only of the projected emittance $\bar{\varepsilon}_x$ or $\bar{\varepsilon}_y$. For systems with axial symmetry 'radial emittance' is sometimes used, with definition similar to Eq. 2 but with r in place of x. In a linear system this appears quite logical, since the envelope equation is as Eq. 2 with r substituted for x and y. This is convenient for examining non-linear systems in which some form of spherical aberration is present, such as might arise from imperfect lenses or non-linear space-charge forces in a beam with non-uniform radial density distribution. (As an example see ref. 6, this conference). It must be emphasized, however, that in general, although $x^2 + y^2 = r^2$, it is not true that $x'^2 + y'^2 = r'^2$ unless the initial beam has no particles with angular velocity about the axis. The rr' emittance is thus useful for initially laminar beams, but is less so, for example, for beams where thermal velocities arising from cathodes or plasma ion sources are important.

Alternative definitions of emittance have been used including both x and y motion in cylindrical beams. Using a hydrodynamic approach rather than the usual optical one, Lee, Yu and Barletta[7] define it as

$$\bar{\varepsilon}^2 = \bar{r}^2 \left(\overline{x'^2 + y'^2} - \bar{r}'^2 \right) \qquad (4)$$

and use this to calculate emittance growth in a converging beam where the deviation from ballistic trajectories arising from space-charge is small.

THE PROBLEM OF EMITTANCE GROWTH

Of great technical importance is the question of how the emittance of a beam grows as it passes down an accelerator or a beam transport system. Much progress has been made in this field since pioneering work at CERN and Brookhaven over 20 years ago, but the subject is far from completely understood. Modern computers have enabled many impressive computations and similations to be made, but it is still not clear whether any further physical insights await revelation. It is not the intention to review this very extensive work here, but merely to list some questions which seem not to be resolved. It is assumed that the reader is familiar with recent work of Hofmann, Reiser, Wangler and others and earlier work of Lapostolle concerning the concept of non-linear field energy and its conversion to particle energy in the process of emittance growth. (See for example ref. 8 and earlier references therein). Some questions to which there do not seem to be agreed answers are listed below. Some of these will be discussed further in the next section.

(1) A beam is fed into a long uniform focusing system. In the presence of space-charge, but not Coulomb collisions, is a final equilibrium with $v_z \gg v_x, v_y$ ultimately attained?

(2) If so, is it possible other than by direct computer simulation to determine the final emittance $\bar{\varepsilon}_f$ from the initial value $\bar{\varepsilon}_i$ and the initial matching conditions?

(3) Does the velocity distribution eventually become Maxwellian in the absence of collisions?

(4) Can we make any estimate of the time to reach equilibrium in terms of ε_i, σ/σ_0 and matching conditions? (Here σ/σ_0 is the tune depression of the 'equivalent' K-V beam with the same value of $\bar{\varepsilon}$).

(5) What happens in a periodic system? Does the beam evolve to a periodically fluctuating state with Maxwellian velocity distribution?

(6) In a beam with unequal transverse energies in the two directions parallel to the symmetry planes, under what conditions is equipartition achieved? What is the physical mechanism?

(7) As 6, but for bunches in a linac.

(8) Can we find a way of estimating these equipartition times?

It is not difficult to add further related questions to this list. In the next section the physical nature of emittance is discussed, in the hope that this may lead to some further insights into the above questions.

THE PHYSICAL NATURE OF EMITTANCE

Emittance may be regarded as a figure of merit for the quality of a particle beam; it is clearly related to brightness, a fundamental concept in light optics, and much discussed also in electron optics. (A typical paper in which these relationships may be seen is given as ref. 9). Although brightness is strictly defined locally, the overall brightness of a particle beam is often defined as proportional to the current divided by the product of ε_x and ε_y. In both light optics and charged-particle optics there is a tendency for the quality of a transmitted beam to be degraded by aberrations and by misalignments and distortions of the focusing elements. In charged-particle optics there is the additional complication of self-fields.

An alternative point of view is to regard the emittance as representing the effect of a force tending to disperse the beam. The envelope equation (3) can equally well be expressed in terms of time as the independent variable. Multiplying Eq. 3 by term by $\gamma m_0 \dot{z}^2$ converts it into $\gamma m_0 \ddot{x}$. If we now consider the force on a small volume element of the beam rather than on individual particles as before, the second term represents the attractive focusing force, and the fourth the space charge repulsion. It is not difficult to show that the emittance term arises from a negative radial pressure gradient[10]. For a matched gaussian beam the temperature is everywhere constant, but the density and hence nkT decreases with radius. For a beam with a K-V distribution on the other hand the density is constant but the pressure decreases with radius.

Perhaps the most fundamental viewpoint, however, is obtainable from statistical mechanics. The particles in the beam represent an ensemble in phase-space evolving in time under the constraints of Liouville's theorem. As observed earlier, the fundamental constraint is that the phase-space density in the neighbourhood of a particular point remains constant. For a given distribution it is by no means immediately evident how to use this fact to determine how the various moments of the distribution, such as those specifying the r.m.s. emittance or beam radius, evolve with time. Energy must also be conserved, and it is now believed that in an external potential independent of z the <u>transverse</u> energy (kinetic + potential + field energy) observed in the beam frame is conserved. This enables correlations between radial density distribution and emittance to be made, but does not allow predictions of how either of these quantities varies with z.

At this point we make connection with the optical viewpoint and note that aberrations and non-linearities in a focusing channel, act on a beam that is not matched, (so that the distribution is not independent of z), to cause phase mixing of the osillations of individual particles, and filamentation of the phase space. Filamentation can also occur in a drifting expanding beam. Such filamentation in general increases the projected emittances of the beam. (This is not always so, artificial singular distributions can be constructed in which $\bar{\varepsilon}$ decreases monotonically[11]). Coupling, even in linear systems can cause 'twisting' of the phase-space distribution in which the values of the projected emittances ε_x and ε_y oscillate. This is observed in the well known n = 0.2 coupling resonance in cyclotrons.

EMITTANCE AND ENTROPY

A connection between emittance and entropy was first noted as long as 30 years ago. This idea was later explored in a short paper, and an attempt was made to relate the entropy to other variables regarding the beam as a 'drifting gas' described in terms of thermodynamic variables such as pressure, temperature, and internal energy[12]. This did not lead to any new insights of practical value, and the connection has perhaps acquired the status of an intellectual curiosity. It is worth re-examining the question, to see whether any useful results might be obtained. For simplicity, we start as in ref. 12 with a one dimensional system described by an ensemble of N points on the xx' plane. This plane is then divided into cells of area A, such that each cell contains many points, but the cell size is small compared with that of the overall distribution. With these assumptions, and Boltzmann's definition of entropy,

$$S = k \ln W, \qquad (5)$$

where W is the number of ways that the points can be assigned to the cells to produce the given distribution, it is shown in ref. 12 that for a uniformly filled ellipse of area $\pi\varepsilon$

$$S_0 = \ln \pi\bar{\varepsilon} - \ln A, \qquad (6)$$

where S_0 has been written for S/kN. (Note that $\bar{\varepsilon}$ is the r.m.s. emittance including the factor 16 in Eq. 3). For other distributions there is a different numerical factor under the logarithm in the first term on the r.h.s. of Eq. 6, leading to an additional constant to be added to S_0 in Eq. 6.

One feature that this model shows is the growth of effective emittance arising from filamentation. If the density distribution varies by only a small amount from one cell to the next, distortion of the shape of

8 The Emittance Concept

the distribution does not affect the entropy. If, however, the distribution becomes excessively filamented this is no longer the case. When the width of the filaments becomes comparable or less than the cell size more cells are occupied, and the entropy increases. If the cell size is set equal to that which can be resolved by the measuring apparatus, then the emittance increase becomes analogous to the entropy increase of two almost identical fluids that are mixed. This is the classic 'mixing' problem. If some red fluid is dropped in to an otherwise identical colourless fluid, the boundary between the two can at first clearly be seen; later mixing occurs until the whole gradually assumes a uniform pink colour.

Following the analogy between emittance and entropy, both would at first sight appear to be a measure of the disorder of the system. Unfortunately the correspondence is not so good as might first appear. For example, returning to the one dimensional system described above, if instead of the points being uniformly distributed within an elliptical contour they are distributed round its circumference only, (corresponding to particles all oscillating with the same amplitude but uniformly distributed in phase,) the entropy and emittance are very different. Because of the very small number of cells occupied the entropy will be extremely small. The r.m.s. emittance will, on the other hand, be four times as large. Likewise, for a two dimensional system the K-V distribution is represented in phase-space by a three-dimensional shell and thus has zero four-dimensional hyper-volume. This represents a highly ordered system with low entropy. The less ordered 'waterbag', in which the centre of the hyperellipsoidal shell is uniformly filled, has very much higher entropy. If, on the other hand, projections of these two distributions on the xx' plane are considered, the ellipse is filled, uniformly for the K-V distribution and with parabolic density profile decreasing to zero for the waterbag; the emittance of the less ordered waterbag distribution is now lower.

A further example to illustrate this is the projected emittance of a uniform beam of radius a_0 with axial symmetry confined by electrostatic lenses, but rotating with angular velocity $\dot{\theta}$ about the axis. The flow is laminar and therefore highly ordered, the temperature is zero, but the projected emittance on the xx' plane is an ellipse of area/π given by

$$\varepsilon = a_0^2 \, \dot{\theta}/v_2. \tag{7}$$

In xx' yy' space the points corresponding to this distribution lie on a two-dimensional surface with entropy even less than that of a K-V distribution with the same projected emittance. (Note, however, that for a rotating beam confined by a solenoid, the emittance must be measured in the

Larmor frame. It may then be zero if $\dot{\theta}$ is zero in this frame, as would be the case, for example, with Brillouin flow).

We conclude that emittance cannot be directly related to disorder. Despite these anomalies, the correspondence between emittance and entropy might not be so bad for less singular distributions. In many practical situations the transverse velocity distribution is not far from Maxwellian. Consideration of the entropy might enable the final equilibrium configuration of a long beam to be determined, though it must first be established that such an equilibrium exists. To avoid this difficulty consideration of the following simpler problem is proposed. A parabolic potential well with axial symmetry is provided over a finite length. This might be a section of a charged cylinder which is transparent to charged particles. At time t = 0 a gas of charged particles with axial symmetry, radial velocity, and arbitrary radial distribution which is independent of z is released in the potential well. The particles move radially, and because of the non-linearity of the space-charge force with radius, individual particle oscillations of different amplitude will phase-mix. This is similar to, but simpler than, the problem of a mismatched beam launched into a uniform focusing channel. (If the presence of ends is objected to, a ring configuration can be substituted). Now it is reasonable to suppose that this simpler artificial system will arrive at some final equilibrium. The following questions might be asked about its behaviour.

(1) Does the particle motion remain both independent of z and without z-velocity, or does the emergence of chaotic behaviour give rise to z-variation and z-velocities?

(2) If so, does this imply that in a beam there is no final steady state with $v_z \gg v_x, v_y$?

(3) Does the distribution become Maxwellian, and if so is there equipartition between radial and z directions?

In the above questions collisions have been ignored, but we know that they will ultimately lead to a Maxwellian distribution in three dimensions, with conserved energy and maximum entropy. We may further enquire:

(4) Can anything be said about relative time constants for relaxation to a Maxwellian distribution by collisions and the corresponding relaxation time by chaotic effects (if there is one)?

(5) Is the final distribution obtained by solving Poisson's equation with the Boltzmann equation[13], with energy conservation and the

constraint that the entropy should be a maximum? (The self-fields will contribute to the internal energy U of the system).

At this point it may be asked what the relevance of considering the above questions is to the problem of how the emittance and radial distribution of a beam develops. It can only give some indication of what the behaviour after a very long time might be, and this is probably of no practical interest at the present time. Transport systems of such great length do not exist, and if they did then the beam behaviour would probably be more influenced by misalignments of the focusing elements.

More generally, it may be asked whether the concepts of statistical mechanics and thermodynamics can yield new insights, or put on a more firm footing recent work on emittance growth.

CONCLUSION

Understanding of emittance growth and equipartition phenomena in particle beams where space-charge forces are large are incomplete. It is not clear whether any new physical principles remain to be discovered that might provide insights of practical value. It is hoped that the informal discussion in the present paper might suggest lines for further thought and enquiry.

REFERENCES

1. T. Sigurgeirsson, CERN report CERN/T/TS-2, (Dec 1952).
2. P. M. Lapostolle, IEEE Trans. Nucl. Sci. **NS-18**, 1101 (1971).
3. F. J. Sacherer, IEEE Trans. Nucl. Sci. **NS-18**, 1105 (1971).
4. J. D. Lawson, The Physics of Charged Particle Beams, 2nd edition. (Clarendon Press, Oxford, 1988).
5. A. J. Dragt, this conference.
6. T. P. Wangler, this conference.
7. E. P. Lee, S. S. Yu and W. A. Barletta. Nuclear Fusion **21**, 961, (1981).
8. T. P. Wangler, K. R. Crandall, R S Mills and M Reiser, IEEE Trans Nucl. Sci. **NS-32**, 2196 (1985).
9. J. P. Davey, Optik **33**, 580 (1971).
10. J. D. Lawson, Plasma Physics **17**, 567 (1975).
11. Ref. 4, Appendix 5.
12. J. D. Lawson, P. M. Lapostolle and R. L. Gluckstern, Particle Accelerators **5**, 61 (1973).
13. Ref. 4, Section 4.6.1.

EARLY STUDIES ON INTENSITY LIMITATIONS IN PROTON LINEAR ACCELERATORS

P. Lapostolle

ABSTRACT

Soon after the successful operation of the first AG proton synchrotrons an effort was made to increase their intensity and the intensity of their injectors: linear accelerators. With the help of elaborate beam dynamics codes, simulation work started, in particular at CERN, including the so-called space charge effects. As will be described, at the same time as computations were made modelling the complex behavior of an accelerated beam, the simpler case of a pure beam transport was studied with the aim of guiding theoretical work. Such early computations, however crude they were at that time – in the late sixties or early seventies, – still remain a useful basis for the more recent elaborate developments.

INTRODUCTION: PRELIMINARY SPACE CHARGE STUDIES

As was well known from the principles of physics and experience from the vacuum electron tubes, the influence of charge repulsion in particle beams is to counteract the focusing action developed for guiding them. If negligible for very low intensities, it progressively increases with it. At the time the first CERN PS and Brookhaven AGS were designed only low intensities were considered (a few mA at injection) and this problem, if not ignored, was only really taken into account in the low energy beam transport, between the source and the linac.

Space charge in a synchrotron considered as a beam channel with a slow acceleration

Already in 1959 Kapchinskij and Vladimirskij[1] had presented their work on space charge effects in a beam transport system and the so-called K-V equations. With a special shell distribution of particles in phase space leading to a uniform distribution in the beam elliptic cross section, space charge field is linear and envelope equations were derived.

These equations were used in the mid 60's for various studies of the synchrotron limitations; this work included, for instance, an extensive study of the envelope oscillation modes (symmetrical and antisymmetrical), their dependence upon beam intensity and emittances and oscillation amplitude, their excitation by a periodic focusing perturbation, and the amplitude induced through a resonance crossing. Such a work performed independently by Sacherer[2] and with even more detail by Lapostolle and Thorndahl[3] gave a very good overview of the properties of envelope oscillations and their mechanism. One interesting result was, for instance, the following observation: as was said, in a K-V beam one has two modes of envelope oscillations, the frequency of which is different in the case of space charge, but other frequencies exist in a beam: apart from pure transverse coherent oscillation frequencies of the beam as a whole (which do not depend upon space charge except through image effects) individual particles perform incoherent oscillations at still other frequencies. The interesting property observed is that when an external perturbation becomes in resonance with the corresponding incoherent frequency the envelope oscillation which is induced modulates the space charge linear field in such a way as to exactly cancel the external perturbation; there is no resonance to be observed with the incoherent oscillation frequencies, but only with the envelope oscillation modes (at least in this linear field case).

Space charge beam dynamics in linacs

The above studies considered a 2 d. problem corresponding to a continuous beam or very long bunches as exist in a synchrotron. In a linac, the bunches, if assumed to be ellipsoidal, have their three axes of similar lengths and the possibility of treating separately transverse and longitudinal motions is not obvious.

A numerical analysis was adopted: an intensive work took place originally at MURA and the program PARMILA was developed using the so-called Panofsky equations for the treatment of acceleration in the gaps and a multiparticle simulation for the space charge. Promé from Saclay, after a stay in MURA, wrote in collaboration with Martini from CERN another code MAPRO making use of the new beam dynamics equations of Lapostolle.[4,5] PARMILA having also adopted the new equations the two codes then gave similar results.

For a given machine they show quite well its properties and its capabilities. They did not help, however, for any detailed understanding of the space charge mechanisms even though several interesting properties were observed.[6] No real guidelines for an improved design of a linac were really yet derived even though some proposals were submitted.[7]

Stationary and non-stationary distribution in phase space

In the case of linear motion any stationary distribution in phase space must be of elliptic shape – or hyperellipsoidal shape when several degrees of freedom are considered.[8]

At the time of these early space charge studies, the only well known stationary distribution for a continuous beam was the so-called Kapchinskij-Vladimirskij distribution[1] already mentioned. It is a shell distribution on a hyperellipsoid which is not very physical (except, for instance, with a special multiturn injection in a synchrotron). Usual beams are more of a bell shape type or maybe, if diaphragms are used, of uniform distribution in phase space (waterbag). Kapchinskij had al-

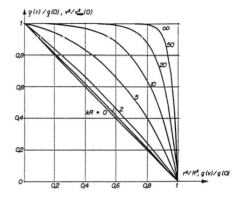

Fig. 1. Density $\rho(r^2)$ as a function of radius r for a waterbag stationary circular distribution of radius R:
$$k^2 = \frac{2q\rho(0)}{\epsilon_0 m v_{\max}^2(0)}.$$

ready treated this last problem in his Russian book[9], but this work was not well known. It was derived again a few years later[10] and then several other distributions, including gaussian, were treated.[11,12,13]

According to the beam brightness and beam radius the shape of the boundary of a stationary waterbag distribution in the 4 d. phase space changes from a hyperellipsoid for a very low density beam to what can be called a bicylinder for an infinite intensity. For a beam of circular symmetry, Figure 1 shows the density distribution as a function of radius. For a very high intensity the density becomes flat with a fast drop in a thickness that Hofmann[14] has shown to be the Debye length. The beam behaves like a plasma where the distribution of the charged particles is such as to the cancel the external focusing forces. Particles then move freely inside the beam and are reflected on its boundaries. The same property is observed for other distributions like gaussian just with the addition of a halo around.

SIMULATION STUDIES

The code used in the early studies at CERN was of the *particle in cell* type (PIC) in order to avoid *collision effects*. Fields were originally computed by summation over the charges present in each cell assuming for space charge a fourfold symmetry.[15] When *fast Fourier transform*[16] (F.F.T.) became available it was used with only a two-fold symmetry assumed. In order to obtain the maximum accuracy a variable mesh was used the size of which exceeded only by 10% the coordinates of the outermost particles. Care was taken to use a beam dynamics integration routine adequate to satisfy Liouville's theorem. A tentative to check Poincaré invariants on the results remained, however, inconclusive.

In these initial computations one difficult choice concerned the way in which to present the results. Density in cross section and emittance plots in phase planes could be made, but numerical values or curves are necessary for any quantitative comparison. Experimental emittance results were at that time at CERN currently

Fig. 2. Emittance-current curves for a stationary waterbag distribution as in Fig. 1. The shape of the curves does not depend much upon the beam current.

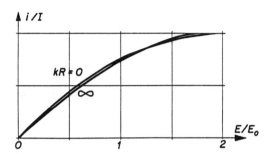

given in the form of an emittance current curve: having drawn in a phase plane equidensity contours and integrated in each of them intensity and area, the set of values obtained allows the plot of a curve as shown in Fig. 2. Such a method was transposed on the computer.

Quasi-stationary circular beams with continuous focusing[17]

When the code was operational the first runs were to check its accuracy by testing the way known distributions were computed. Cross section distributions and emittances proved to be very well conserved for the stationary distributions – K-V, waterbag, gaussian – though some statistical noise appeared around the beam where a small halo was formed.

Distorted distributions were then tested: uniform distribution in a hyperellipsoid or bicylinder, uniform density in configuration space and gaussian in velocity ... with also some inevitable mismatches. Results can be summarized as follows:

> The most apparent phenomenon is a density oscillation in the beam as shown in Fig. 3 – such oscillations were computed by Gluckstern.[18]

> Matching had to be made empirically; no rule was found but energy considerations seemed to be a guideline to follow.

> For bright beams with quasi-uniform density K-V equations seemed to give, using the outer radius, the correct oscillation frequency.

Eventually a test was performed to check the plasma analogy of a bright beam where the density tends to cancel the focusing force inside the beam: with a non-linear focusing field the density distribution around which the beam oscillates is no longer uniform but satisfies the plasma property (see Fig. 4).

Non-circular oscillating beams with continuous focusing[19]

While the plots used above for representing the beam state were very instructive, they were still not very easy to use. Envelope oscillations, for instance, could

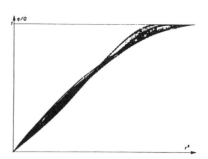

Fig. 3. Oscillations of the charge distribution as a function of radius (squared) for an initially non-stationary distribution (Gluckstern modes).

not be observed clearly due to the necessarily limited number of plots taken. The presence of a halo, as mentioned, made arbitrary the value of an outer axis. So was is decided to use mean square values, meaningful for any distribution (gaussian as well). For emittance the choice of a skew axis of slope α – looking for a minimum with respect to α – allows the computation of an emittance area which turns out to be the second order invariant of a linear motion.[20]

In order to be consistent with usual K-V expressions which refer to the uniform distribution case, it was decided to introduce *effective beam dimensions and emittances* as:

$$a = 2\sqrt{\overline{x^2}} \qquad b = 2\sqrt{\overline{y^2}}$$

$$\epsilon = 4\sqrt{\overline{x^2}\,\overline{x'^2} - \overline{xx'}^2}.$$

With these definitions the following observations could be made:

For symmetrical and antisymmetrical oscillations $\overline{x^2}$ and $\overline{y^2}$ perform quasi-sinusoidal oscillations (their sum is almost constant for antisymmetrical oscillations).

For small enough amplitudes and not too large brightness the amplitude is practically constant and ϵ, apart from an initial change which depends upon the density distribution, also remains constant.

For large amplitudes and not zero intensity the still sinusoidal oscillations are damped but the average remains practically constant. In this case ϵ goes up (Fig. 5).

The most important result, however, was that, from the observation of many cases, K-V equations are always satisfied locally by effective values a, b, and ϵ. Such a property was later proven by Gluckstern[21] and Sacherer.[22]

This property allows the computation (to replace the previous empirical process) of matching conditions for any stationary distribution and of initial conditions for the excitation of pure modes of envelope oscillations. This observation was in fact one of the clues to discovering the relevance of the above generalized K-V equations to any particle distribution.

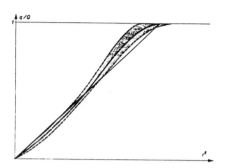

Fig. 4. Charge distribution inside a beam for a nonlinear focusing system: Space charge tends to cancel the external field.

An experiment was also made to observe the possible instability of the Gluckstern modes for a K-V distribution. The interesting result is that nothing is seen on a, b, or ϵ nor much on the emittance or distribution plots but a check of the thickness of the 4 d. K-V shell clearly showed the instability when it is present (see also Hofmann[23]).

Alternating gradient focusing[24]

When the focusing strength is equal in the two transverse directions and emittances are also equal, K-V equations can be normalized to a simple form like

$$u'' + u - \frac{1}{u^3} - \frac{2\delta}{u+v} = 0$$

$$v'' + v - \frac{1}{v^3} - \frac{2\delta}{u+v} = 0.$$

The space charge parameter δ can be written in the non-relativistic case (the most critical for linacs) in the form:

$$\delta \simeq \frac{1}{200} \frac{\lambda_\beta I}{V \epsilon_n}$$

where λ_β is the betatron wavelength in meters, I is the electrical intensity in mA, V is the voltage in MV which would be necessary to bring the particles to their actual velocity, and ϵ_n the normalized effective emittance (area divided by π) in mm-mrad (such an expression is also valid for any multicharged particle).

To refer to usual practice one can note that

$$\delta = \frac{\sigma_0}{\sigma} - \frac{\sigma}{\sigma_0}.$$

Simulation of A.G. focusing was made for simplicity with an F.D. structure. Four values of δ were taken:

0.4   1.5   5.5   22

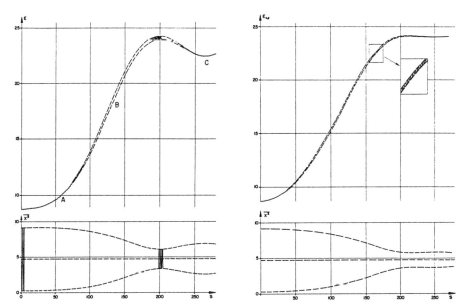

Figure 5: Evolution along a continuous focusing transport system of a beam performing symmetrical (left) and antisymmetrical (right) envelope oscillations; the curves show the emittance growth and the mean square axes oscillations.

and four values of σ_0:

$$40° \quad 75° \quad 115° \quad 150°$$

in order to examine the effect of these parameters on the beam stability and its sensitivity to resonances.

The results can be summarized as follows:

> For small enough δ (smaller than 0.5, for instance) emittances are conserved for any σ_0 and the halos formed do not extend very far (apart, of course, when lying on a resonance).

> For large δ resonance appear when $\sigma_0 > 90°$ to an extent which seems to grow with δ as well as with σ_0.

> For small σ_0 (less than 60°) δ can reach very high values ($\sigma/\sigma_0 = 0.05$) with only a very slow emittance increase (which Haber[25] has shown later to be a pure numerical effect due to some statistical noise).

A few runs were made with equal focusing strengths in x and y but unequal emittances: one observes a relatively fast transfer of emittances with equipartitioning (the speed increases with δ). Other runs were made with equal emittances and unequal focusing strengths: if the focusing strengths are different enough nothing happens to otherwise stable beams.

TOWARDS SOME THEORY: CONCLUSION

One purpose of the simulation was to guide some theory for a better understanding of space charge phenomena in intense beams of large δ. The properties of K-V equations and some remarks on energy, induced such a theoretical effort. Following the initial proof of Gluckstern[18] a derivation was made for a non-circular problem. While Sacherer[22] treated rigorously the elliptic case, Lapostolle[26] tried to develop a theory for a more general situation (simulation shows that the beams doe not remain necessarily elliptical). Corrections with respect to ellipticity remained small but were suspected to be perhaps large enough to be responsible for the emittance growths observed with periodic A.G. focusing in very bright beams of large δ; in fact it was later shown that similar growths also occur for circular beams and solenoid focusing so that only resonances are responsible and not any departure from ellipticity (see among other Reiser et al.[27]).

Nevertheless, a new derivation of generalized K-V equations was made from energetic considerations[26] leading in particular to an equation of the emittance growth resulting from a change in particle distribution. Such a work was developed later by Wangler et al.[28] and extended to the 3 d. case of bunches by Hofmann.[29] Anderson[30] gave a more detailed analysis of the way such a redistribution takes place showing that the speed of it is related to the plasma frequency of the beam.

Energy conservation in a beam with invariance of its emittance under some reversible circumstances led also to a comparison with thermodynamics, where entropy replaces emittance.[31] Thermalization effects clearly observed in simulation – and of course in real beams[32,33] – could not, however, be derived theoretically even if nothing objected to it. It is only later that Jameson[34], assuming that the phenomenon was taking place and knowing how damaging it could be for the beam brightness, derived equations for the design of linacs with very little emittance growth (an approach of a similar nature as the one used by Kapchinskij for his R.F.Q.s with minimal emittance). Such a work was still extended later by Wangler et al.[35]

As can be seen, despite several misleading ideas and sometimes unnecessary efforts, most of the concepts derived from the early computation and simulation work did help to develop the present theory of bright beams. If a few questions are not yet solved theoretically the possibility exists today to design good linacs producing intense beams of high brightness.

REFERENCES

1. I. Kapchinskij and V. Vladimirskij, Proc. H.E. Accel. Conf., Geneva (1959), pp. 274-288.
2. F. Sacherer, Ph.D. Thesis, L.R.L. Rep. UCRL 18454 (1968).
3. P. Lapostolle and L. Thorndahl, Part. Accel. Conf., Washington, IEEE Trans. **NS-14** (1967), pp. 567-571.
4. P. Lapostolle, CERN Rep. AR/Int. SG5/11 (1965), and LINAC Conf. LA 3609, Los Alamos (1966), p. 201.
5. P. Lapostolle, CERN Rep. AR/Int. SG/65-15 (1965).
6. M. Promé, These no 761 Orsay, LASL Trans. LA-TR-79-33 (1971).
7. P. Lapostolle, CERN Rep. ISR-300 LIN/66-33 (1966).
8. P. Lapostolle, CERN Rep. ISR-300 LIN/66-32 (1966).
9. I. Kapchinskij, LASL.LA TR 80-10 (1980); also *Theory of Resonance Linear Accelerators* (Harwood Publishers, 1985), see §3.3.
10. P. Lapostolle, Proc. H.E. Accel. Conf., Erivan (Acad. Sci. Armenian SSR, Erivan, 1970), p. 205.
11. M. Promé, Saclay Rep. SEFS TD 69/36 (1969).
12. R. Gluckstern et al., Proc. LINAC Conf., Batavia (1970), pp. 823-850.
13. F. Sacherer, CERN Rep. SI/Int. DL/70-5 (1970).
14. I. Hofmann, "Transport of high intensity beams," in sl Advances in Electronics and Electron Physics, Supplement C: Applied Charged Particle Optics, (1986) p. 63.
15. P. Tanguy, Proc. LINAC Conf., Batavia (1970), pp. 771-790.
16. P. Lapostolle and B. Lapostolle, CERN Rep. ISR-300/ LI/69-43 (1969).
17. R. Gluckstern, Proc. LINAC Conf., Batavia (N.A.L., Batavia 1971), pp. 811-822.
18. P. Lapostolle, CERN Rep. ISR/DI-70-36; also Part. Accel. Conf., Chicago, IEEE Trans. **NS-18**, 3 (1971) pp. 1101-1104.
19. L.C. Teng, NAL Rep. FN-221-0100 (1971).
20. R. Gluckstern, (private communication, 1970), correspondance and discussion during the LINAC Conf., Batavia, 1970.
21. F. Sacherer, Part. Accel. Conf., Chicago, IEEE Trans. **NS-18**, 3 (1971), pp. 1105-1107; see also CERN Rep. SI/Int. DL/70-12.
22. I. Hofmann, Part. Accel. Conf., Washington, IEEE Trans. **NS-28** (1981),pp. 2399-2401.
23. P. Lapostolle, CERN Rep. ISR/78-13, (1978).
24. I. Haber, A.I.P. Conf. Proc. 139 (1986).
25. P. Lapostolle, CERN Rep. ISR-DI/71-6, LANL Trans. LA-TR-80-8, (1971).
26. M. Reiser et al., Part. Accel. **14** (1984), pp. 227-260.
27. T. Wangler, M. Reiser, et al., Part. Accel. Conf., Vancouver, IEEE Trans. **NS-32**, 5 (1985), pp. 2196-2200; also HIF Conf., Washington, AIP Conf. Proc. **152** (1986), pp. 166-185.

28. I Hofmann, Proc. LINAC Conf., SLAC (1986) pp. 183-187.

29. O.A. Anderson, HIF Conf., Washington, AIP Conf. Proc. **152**, pp. 253-263; also Part. Accel. **21** (1986), pp. 197-226.

30. J. Lawson, R. Gluckstern, and P. Lapostolle, Part. Accel. **5** (1973), pp. 61-65.

31. R. Chasman, BNL. LINAC Conf., BNL Rep. 50120 (1068) pp. 372-377.

32. P. Lapostolle et al., CERN Rep. 68-35, (1968).

33. R. Jameson, Proc. LINAC Conf., Santa FDe (1981), pp. 125-129.

34. T. Wangler, I. Hofmann, and F. Guy, Proc. LINAC Conf., SLAC Rep. 303 (1986), pp. 340-345.

EMITTANCE GROWTH FROM SPACE-CHARGE FORCES

by

THOMAS P. WANGLER

Accelerator Technology Division, MS-H817
Los Alamos, NM 87545

Space-charge-induced emittance growth has become a topic of much recent interest for designing the low-velocity sections of high-intensity, high-brightness accelerators and beam-transport channels. In this paper we review the properties of the space-charge force, and discuss the concepts of matching, space-charge and emittance-dominated beams, and equilibrium beams and their characteristics. This is followed by a survey of some of the work over the past 25 years to identify the mechanisms of this emittance growth in both ion and electron accelerators. We summarize the overall results in terms of four distinct mechanisms whose characteristics we describe. Finally, we show numerical simulation results for the evolution of initial rms-mismatched laminar beams. The examples show that for space-charge dominated beams, the nonlinear space-charge forces produce a highly choatic filamentation pattern, which in projection to the 2-D phase spaces results in a 2-component beam consisting of an inner core and a diffuse outer halo. In the examples we have studied the halo contains only a few percent of the particles, but contributes about half of the emittance growth.

1. INTRODUCTION

Many accelerator applications require output beams with high phase-space density or high brightness. To achieve this goal it is necessary to control all sources that cause dilution of the phase-space density. In practice, it is difficult to measure the full six-dimensional phase-space density. Instead, the projected two-dimensional phase-space distributions are measured, and the effective areas occupied by the beam in those projections are characterized by rms emittances, which can be calculated for any arbitrary distribution. The evolution of the rms beam size is expressed in terms of rms emittance through the envelope equation.

In the presence of nonlinear forces or coupling between planes, arising either from external focusing or from self fields, the rms emittances can increase even when Liouville's theorem is satisfied. In general, for those applications that require an output beam capable of being focused to a very small spot, such emittance growth effects must be avoided. Examples include intense heavy ion beams for heavy-ion fusion and high-brightness photocathode electron guns for electron linear accelerators. These phenomena are studied using computer simulation of the multiparticle dynamics.

For beams with high average intensity, one may be concerned not with the rms or average phase-space areas, but with the outer part of the distribution, which determines particle losses on the accelerator structure. Relatively small losses in a high-energy accelerator may produce enough radioactivation of the accelerator structure or radiation damage of components to create practical difficulties in maintenance and operation of a facility.[1] For this case even if the focusing of an intense output beam does not impose difficult requirements on the final emittances, rms-emittance growth is still to be avoided because such growth generally means that the population of the outer regions of phase space has increased, an effect known as beam-halo formation. When this is a concern, the use of a single number (such as rms emittance) to characterize the distribution has limited usefulness, and one must look in more detail at the distribution.

A major cause of emittance growth in low-velocity intense beams is the Coulomb self force. In most accelerator beams this is predominantly a collective force, and small-impact-parameter binary collisions are usually believed to have little effect on the dynamics. This smoothed or average Coulomb force is called the space-charge force and is described by a repulsive self-electric field and an attractive self-magnetic field. The magnetic term is only important for relativistic beams and its contribution reduces the total space-charge force. In recent years emittance growth mechanisms from space-charge forces have been studied by computer simulation of intense low-velocity beams, especially for ion linear accelerators and beam transport lines. This has resulted in increased understanding of the mechanisms of this rms-emittance growth. In circular accelerators the space-charge force causes a spread of the betatron frequencies, which, in the presence of nonlinear resonances caused by magnetic field errors, also leads to emittance growth.[2] In this paper, however, we will restrict the discussion to emittance growth caused directly from the space-charge forces.

The concept of emittance and the definition of rms-emittance have been reviewed by Lawson.[3] It should be noted that the definition in this paper differs from his by not using the factor of 4 for the rms emittance. The K-V envelope equation is also discussed in Lawson's article, in which it is pointed out that the emittance term corresponds to a negative radial pressure gradient, which when added to the space-charge term gives the total effective repulsive force that affects the rms beam size. Comparison of the space-charge and emittance terms establishes the general criterion for determining the conditions under which space-charge effects are large enough to be of concern. Thus, when the ratio of the space-charge to emittance term approaches or exceeds unity, the space-charge force will generally be important.

The space-charge force is complicated because the field depends upon the time-varying charge density of the beam. In general, it is nonlinear and time dependent, and one observes coupling between the three planes. In the presence of external focusing forces, one observes phenomena that are common in plasma physics, such as plasma oscillations and Debye shielding. The plasma period determines a basic time scale for these phenomena, and the Debye length determines a basic length scale for the particle distribution. The net force, consisting of the external focusing plus the time-dependent space-charge force, may be either attractive or repulsive, and the sign of the net force may even vary across the beam. These conditions can lead to very chaotic behavior, as will be discussed later, and one must rely on numerical simulation codes to study the detailed dynamics.

2. EMITTANCE GROWTH MECHANISMS

2.1. Matched Beams

We distinguish between what we will call internal matching and rms matching. Internal matching constrains the six-dimensional phase-space distribution to make the isodensity contours coincide with the particle phase-space trajectories. For an internally matched beam, the distribution will be in equilibrium in the accelerator channel, and no emittance growth will occur, even though nonlinear forces may act on the beam. Such an equilibrium distribution is independent of time (stationary) if the focusing is uniform along the accelerator, or is periodic for a periodic-focusing channel. Examples of equilibrium distributions have been studied for two-dimensional transport channels.[4] The most frequently studied is the Kapchinskij-Vladmiriskij (K-V) distribution.[5] Unfortunately this distribution is physically unrealistic because the beam is distributed on the surface of a hyperellipsoid in four-dimensional phase space, resulting in no particles in the central core of this four-dimensional space. This distribution results in uniform ellipses for all two-dimensional projections.

Given a beam that is not internally matched one would like to be able to transform it into an internally matched equilibrium distribution, without increasing the rms emittance in the process. We do not know whether it is possible in principle to perform such a transformation without accompanying emittance growth. Nevertheless, it is feasible to match the rms beam sizes for each degree of freedom. This is accomplished by providing a beam-optics transformation to make the rms sizes constant. In a periodic channel the rms sizes will undergo a periodic flutter about their average values. An rms-matched beam is not generally internally matched, and beam distribution is not in equilibrium; therefore, the beam will evolve with the possibility of irreversible emittance growth. Nevertheless, rms matching is an important characteristic of an internally matched equilibrium distribution, and injection of an rms-matched beam can be considered a first approximation of the desired internally matched beam.

Numerical simulations of nonequilibrium beams in uniform focusing channels show that such beams often evolve to quasi equilibrium distributions, which change only slowly as the beam is accelerated. The evolution of the beams is usually accompanied by rms-emittance change as a result of both nonlinearity and coupling between degrees of freedom. Experience has shown that the velocity distributions of the final beams are Maxwellian-like. When focusing is linear, the spatial distribution of a space-charge-dominated beam consists of an approximately uniform charge-density core of density n. The density increases to zero over a finite distance approximately equal to the Debye length λ_D given nonrelativistically by

$$\lambda_D = \sqrt{\varepsilon_0 kT/nq^2} \; , \tag{1}$$

where q is the charge per particle and ε_0 is the free space permittivity. In Eq. 1 the thermal energy is given by $kT = mc^2\varepsilon^2/a^2$, where mc^2 is the particle rest energy, a is the rms beam size, and ε is the rms-normalized emittance, defined in the Sacherer[6] convention, with no factor of 4 included (but with the relativistic $\beta\gamma$ factor).

For space-charge dominated beams $\lambda_D \ll a$, the equilibrium spatial distribution is approximately uniform with a sharp falloff at the edges. For emittance-dominated beams $\lambda_D \gg a$, the Debye tail occupies essentially the entire spatial extent of the beam, resulting in a peaked Gaussian-like charge density. Among two-dimensional continuous equilibrium beams, the K-V distribution is anomalous because it always has uniform charge density, regardless of the relative importance of emittance and space charge. However, this distribution does not correspond to the final equilibrium state of beams observed in numerical simulation studies.

It is further observed in numerical simulation that the emittance growth of beams that evolve to a final equilibrium distribution is associated mostly with a halo of low-density particles in phase space. This halo is especially undesirable for high-duty-factor linacs because it results in particle losses on the accelerator walls and in radioactivation of the accelerator.

2.2. Historical View of Space-Charge-Induced Emittance Growth

In early emittance growth studies[7,8] at Brookhaven of bunched beams in high-current proton drift-tube linear accelerators, space-charge forces associated with longitudinal to transverse coupling were identified as the primary source of observed transverse emittance growth. It was also found that this emittance-growth mechanism leads to a lower limit for the output emittance as input emittance is decreased at fixed-beam current. Later studies[9] showed that emittance growth could be physically correlated with the dependence of the transverse oscillation frequency on the longitudinal position of the particles in the bunch. At least in the early stages of the emittance growth, the ellipse orientations in transverse phase space depended on the axial position along the bunch, and the overall phase-space area was enlarged.

Early work was also carried out by P. Lapostolle,[10] who combined the numerical simulation studies with analytical work leading to the rms envelope equation and also first described some of the mechanisms discussed in this paper. An observation of longitudinal emittance decrease associated with the transverse emittance increase led Lapostolle to the suggestion of equipartioning.[11] In this picture, the emittance changes are the result of the evolution of a high-current beam towards a thermal equilibrium distribution in which approximate thermal-energy balance or equipartioning, would be established. We refer to this process as the thermal-energy-transfer mechanism. This suggests that if the input beam could be equipartioned in the accelerator in addition to being rms matched, a better approximation to the ideal internally matched beam would result, and therefore space-charge-induced emittance growth would be minimized.

Exact equipartitioning would mean that the mean-square, center-of-momentum velocities in each degree of freedom would be equal. This corresponds to the condition that

$$\sigma_x \varepsilon_x = \sigma_y \varepsilon_y = \sigma_z \varepsilon_z \qquad (2)$$

where σ_x, σ_y, and σ_z are phase advances per focusing period, associated with both the external focusing and the space-charge forces. For a given set of input rms-normalized emittances ε_x, ε_y, and ε_z, and a given beam intensity, the condition given by Eq. 2 imposes a constraint on the relative focusing forces. Thus, while rms matching is achieved by providing a suitable beam-optical matching section before the beam is injected into the accelerator, the equipartitioning condition depends both on the input beam, through the beam current and the emittances, and on the focusing in the accelerator.

Additional understanding of the equipartitioning dynamics was obtained from the work of Hofmann,[12] who identified the longitudinal-to-transverse space-charge effects with coherent instabilities associated with anisotropy in the beam. Underlying this approach is a particular mechanism for the emittance growth;[13] the growth is the result of the excitation of unstable collective modes of oscillation of the beam. Some modes involve radial and azimuthal (quadrupole, sextupole, etc.) density oscillations of the beam. The first reported study of such modes for a two-dimensional, round K-V distribution in a uniform focusing channel was made by Gluckstern,[14] who identified many modes and derived their stability characteristics. Studies for the K-V beam in a quadrupole channel were made by Hofmann et al.[15] and some modes were found to be unstable although not all cases lead to emittance growth.

Hofmann[12] also studied the K-V distribution with an asymmetry between the x- and y-planes and derived the instability thresholds for the different modes. Although the studies correspond to continuous beams in an x-y geometry, Hofmann found that the same instability thresholds were approximately valid for the r-z geometry of a two-dimensional, bunched beam. It was found that equipartitioned beams were stable with respect to these instabilities and that generally the requirement for avoiding emittance growth even allowed some relaxation of exact equipartitioning. The predictions of Hofmann's model were further tested for high-current beams in drift-tube linear accelerators by Jameson[16] who confirmed that equipartitioned input beams produce the minimum emittance growth. Non equipartitioned beams could produce a significant transfer of energy and emittance between the longitudinal and transverse planes. Jameson showed from simulation how the parameters of the nonequipartitioned accelerator beam can change in the space defined by the mode-stability plots derived by Hofmann.[12] This behavior can be complicated and makes it difficult to derive simple design guidelines for avoiding this emittance growth. Perhaps the simplest design approach is to require exact equipartitioning as defined by Eq. 2. A less restrictive guideline is suggested by Hofmann,[12] whose criterion is that energy anisotropy is generally tolerable when the phase advance ratio $\sigma_l/\sigma_t < 1$ where σ_l and σ_t are the phase advances for longitudinal and transverse motion. The growth times from numerical simulation were typically about one to two transverse oscillation periods.

Additional studies of thermal-energy transfer and equipartitioning have been carried out for two-dimensional beams with different initial charge distributions in uniform transport channels.[17,18] Formulas for emittance growth were derived from the relationship between field energy and rms emittance described below, and the formulas were compared with numerical simulations.

A different space-charge-induced emittance-growth mechanism was discovered[19,20] that even affects bunched and continuous beams that are both rms matched and equipartitioned, but are internally mismatched. This mechanism has been called charge redistribution. When a beam is injected into a transport or accelerator channel, the charged beam particles, behaving like a plasma, adjust their positions to shield the external fields from the interior of the beam. For linear external fields in the extreme space-charge (zero-emittance) limit, this implies a charge rearrangement to a uniform density for producing a linear space-charge field with exact shielding. Finite-emittance rms-matched beams in numerical simulation evolve towards an internally matched charge density with a central uniform core and a finite thickness boundary, whose width is about equal to the Debye length. The rms-emittance growth results from the nonlinear space-charge fields, while the beam has nonuniform density and is undergoing internal plasma oscillations. The emittance growth can also be described as the result of the decoherence of the plasma oscillation phases for particles with different amplitudes (phase mixing), resulting in a distortion of the phase-space area. This mechanism of emittance growth has the smallest-known growth time; the full emittance growth occurs during only one-quarter of a plasma period, followed by damped emittance oscillations for typically ten or so additional plasma periods. In a high-current accelerator the full growth can occur within a single cell. This mechanism can become important when beams that have been internally matched to a strong focusing channel are injected after rms matching into a weaker focusing channel. In the strong focusing channel, where the matched beam size is small compared to the Debye length, the equilibrium spatial distribution is a strongly peaked, Gaussian-like distribution. In the weak focusing channel the rms beam size is large, and the corresponding equilibrium distribution is nearly uniform.

If the rms-matched input beam has the peaked spatial profile the beam density will change from peaked to nearly uniform in the weak focusing channel, and the corresponding change in the space-charge field energy of the distribution can be used to calculate the emittance growth. This results from the fact that for a fixed rms beam size, the space-charge field energy is minimum for a uniform beam and increases as beams become more nonuniform. The evolution of the beam from peaked to uniform is accompanied by conversion of space-charge field energy to thermal energy, which causes an increase in emittance. The emittance growth for a spherical bunch containing N particles, each with charge q, is obtained from the expression[21]

$$\frac{\varepsilon_f}{\varepsilon_i} = \left[1 + q^2 N a U_{ni}/60\sqrt{5}\pi\varepsilon_0\gamma^3 mc^2\varepsilon_i^2\right]^{1/2} \quad (3)$$

where a is the rms beam size, U_{ni} is the initial, dimensionless, nonlinear field-energy parameter, a function only of the shape of the initial distribution, and ε_i and ε_f are the initial and final rms-normalized emittances. Emittance growth from charge redistribution is sensitive to the initial spatial charge density. For an initial Gaussian profile of a spherical bunch, $U_{ni} = 0.308$, whereas for a uniform density, $U_{ni} = 0.0$. The equation shows that for a given available field energy, determined by U_{ni}, the emittance growth increases with increasing rms beam size a. This is because emittance is a measure of area occupied in the beam phase space, and for larger beams the increased velocity or divergence spread to be distributed over a larger area.

To avoid emittance growth from charge redistribution, it is necessary either to avoid transitions to accelerator channels with weaker focusing or to always provide input beams with spatial profiles that are as uniform as possible. Other guidelines for minimizing emittance growth from charge redistribution can be inferred from Eq. 3. For a given beam current I, defined as the average value over an rf period, the number N of particles per bunch is given by $N = I\lambda/qc$, where λ is the rf wavelength. Equation 3 predicts that the emittance growth of a bunch is less at high frequencies, a result that appears because a high-frequency linear accelerator has less charge per bunch for a given current. This condition was also reported[9] in studies of emittance growth that included equipartitioning effects and so is more generally valid than for the charge-redistribution effect alone. The charge redistribution mechanism has also been studied by Anderson,[22] who has derived formulas for the dynamics in the extreme space-charge limit.

Numerical simulation studies of transverse emittance growth in a radiofrequency quadrupole (RFQ) linac have also been reported.[23] The main features are:

1) the emittance growth is predominantly caused by space-charge forces, 2) most of the growth occurs while bunching the beam and so is a strong function of the longitudinal beam size, 3) above a certain current, the growth is weakly dependent on the beam current, 4) the growth is almost independent of the initial distribution, and 5) the final emittance approaches a lower limit as the initial emittance approaches zero at fixed-beam current. The emittance growth in the RFQ bunching section may be a combination of the equipartitioning effect and charge redistribution as the bunching forces increase the peak value of the beam current and drive the beam into a more space-charge dominated regime. A semiempirical emittance growth formula was obtained [23] based on Equation 3, which is in good agreement with the numerical simulation results. This formula shows the advantage of high frequency and strong focusing for control of space-charge-induced emittance growth in the RFQ.

When a charged particle beam that is not internally matched has total transverse energy larger than that of a matched beam, excess or free energy can be transformed to thermal energy, resulting in emittance growth, provided nonlinear forces act on the beam. An example is the case of an rms-mismatched beam, where excess potential energy associated with the mismatch is available for emittance growth. Emittance growth is expected when the beam under the influence of the nonlinear space-charge force relaxes toward an equilibrium, or internally matched state. For a uniform, continuous, linear focusing channel, where transverse energy is conserved, Reiser[24] has recently derived an equation for the emittance growth of an initially rms-mismatched beam, assuming that all the excess energy from the initial state is converted into the thermal energy of a final matched beam. Numerical simulation studies[25] confirmed the formula and showed that for rms beam-size mismatches of 50% or more, the emittance growth is the result of a large, well-populated halo surrounding the core of the beam. The studies suggest that rms mismatch may be the source of most of the halo observed in high-current beams. Consequently, we identify this cause of emittance growth as the rms-mismatch mechanism.

In general, the change of the rms emittances can also be related to changes in the field energy associated with the self fields through a differential equation,[20,21,26] which shows that nonlinear space-charge fields are associated with the emittance growth. The application of these ideas to several different problems has been described by Hofmann.[27]

In a periodic focusing channel an additional emittance growth is caused by the envelope instability,[15,28] which occurs when the periodic focusing structure excites coherent modes of the beam. Not only does the envelope grow, but the modes with nonuniform density are excited, and the nonlinear fields cause emittance growth. This emittance-growth effect may be called the structure-resonance mechanism, and it can generally be avoided by designing the transport or accelerator channel at a zero-current phase advance per focusing period no larger than $\sigma_o = \pi/2$.

Space-charge-induced emittance growth also can be important for intense, low-emittance injectors for electron linear accelerators. Much recent work has been motivated by the development of high-brightness photocathode injectors for free electron laser (FEL) applications. For this application, a short laser pulse irradiates a photocathode inside an rf cavity to produce an intense short-bunch that is rapidly accelerated to relativistic energies. Before the electrons reach sufficiently high velocities, for the self-magnetic fields to cancel the self-electric forces, significant emittance growth can occur. As was found in studies of high-current proton linear accelerators[7-9] (see discussion earlier in this paper), most of the initial emittance growth was found to be caused by variation of the transverse space-charge force along the axial position in the bunch as the beam expands in the cavity.[29] It has been found from numerical simulation that the correlation of transverse ellipse orientation with axial position, which is the cause of the rms-emittance growth, can be removed before longitudinal mixing and thermalization occurs by refocusing the beam with a solenoid lens.[30-33] Although it is well known that rms emittance can decrease[20] as well as increase, the example shows that even for space-charge forces, the rms emittance growth is not necessarily irreversible. Also, because the electron beam is being accelerated in the injector to relativistic energies, the system can be designed to effectively freeze the low emittance configuration at the waist before acceleration to the final energy.

In general, the low mass of the electron means that electron beams can be accelerated very rapidly out of the low-velocity regime where space-charge forces have their most serious consequences. This leads to some differences in the space-charge problem for electrons and ions. Though the electron-emittance growth occurs while the beam is expanding and can be reversed by refocusing, low-velocity ions are generally transported and accelerated in a long periodic or quasi-periodic focusing structure. In both cases focusing can be used to control the emittance-growth effects.

2.3 Summary of Space-Charge-Induced Emittance Growth

The characteristics of the four distinct mechanisms: charge redistribution, rms mismatch, thermal-energy transfer, and structure resonances are given in Table I. All four mechanisms share the same fundamental source of emittance growth, which is the nonlinear part of the space-charge force, including coupling effects such as the dependence of the transverse space-charge fields on the longitudinal coordinate. The mechanisms differ in their source of free energy for emittance growth. The structure-resonance mechanism is the only one that is restricted to periodic focusing channels, but the restriction is mostly a theoretical one because practical focusing channels use discrete lenses and are often periodic or quasi-periodic. (In practice uniform focusing channels are thought of as a smoothed representation of a periodic channel.) The time scales listed for the emittance growth are typical effective values, based on the results of numerical simulation studies. Sometimes these studies show a very complex time dependence,[25] and it is clear that a single time constant is not always adequate to describe the physics. Furthermore, one must take care that binary collision phenomena, which can easily occur in the simulation codes, do not produce emittance growth that masks the collective physics we are trying to model. This topic has been treated in two excellent articles by Haber[34] and Haber and Rudd.[35]

Table I shows a strong sensitivity of emittance growth on distribution for both the charge-redistribution and the structure-resonance mechanisms. This conclusion for the former mechanism is well established, but for the latter case it is based on a study showing that emittance growth from structure resonances could be greatly reduced if the transverse velocity distribution was Gaussian-like with a halo.[36] Finally, the suggested cures for each mechanism are given in the last row. To minimize the effect of the structure resonance, we should also add the possibility of injecting with a Gaussian-like velocity distribution.

TABLE I.
MECHANISMS OF EMITTANCE GROWTH
INDUCED BY SPACE-CHARGE FORCES

	Charge Redistribution	RMS Mismatch	Thermal-Energy Transfer	Structure Resonance
Free-energy source	Nonuniform field energy	Potential energy	Thermal energy in other plane	Directed energy
Focusing system	Uniform and periodic	Uniform and periodic	Uniform and periodic	Periodic
Time scale	$\sim \tau_{plasma}/4$	$\sim 10\tau_{plasma}$	$\sim 10\tau_{plasma}$	$\sim 2\tau_{betatron}$
Distribution function sensitivity	Strong	Weak	Weak	Strong
Emittance growth formulas	Yes	Yes	Yes	No
For minimum growth	Uniform density or internal match	rms match	Equipartition	$\sigma_0 < \frac{\pi}{2}$

T. P. Wangler   31

3. PHASE-SPACE DYNAMICS

3.1. Numerical Simulation

So far we have discussed the phenomenon of rms-emittance growth and have identified four fundamental mechanisms, based upon the sources of free energy. Now we will use numerical simulation to look at the multiparticle dynamics and see what changes occur in phase space when the emittance grows. We will examine what may appear to be a relatively simple case of a round continuous beam in a uniform linear focusing channel with purely radial focusing. This system represents a smooth approximation for beams in quadrupole focusing channels and we expect that phenomena observed in the uniform channel will also be observed in the quadrupole channel. We use a numerical simulation code[37] for these studies in which the radial space-charge forces are calculated from Gauss's Law. Consequently, we are studying the effect of the collective forces acting on each particle and ignoring the small-impact-parameter binary Coulomb collisions. Our computer code has been run with 2000 simulation particles through 56 steps per plasma period, choices that we believe are adequate to represent the main features of the space-charge forces. We have chosen to study the dynamics of an initial rms-mismatched laminar (zero emittance) beam. Laminar beams are idealizations because all real beams have finite emittance. Nevertheless, the laminar beam represents the extreme space-charge limit and allows us to isolate the effects of the space charge.

In each of the following figures we show the distributions of a) the radial or $r - r'$ phase space, b) the projected or $x - x'$ (and $y - y'$) phase space, and c) the $x - y$ beam cross section. We show the $r - r'$ phase space because we expect the dynamics to appear simpler in $r - r'$ space when only radial forces act on a laminar beam. We assign an initial positive radius to all particles, but if during the simulation a particle crosses the axis, we change the sign of the radius.

3.2 Mismatched Uniform Density Laminar Beam

We begin by studying the dynamics of an initial uniform-density laminar beam with zero-velocity spread, which is rms mismatched so that the initial rms beam size is larger than the matched value by a factor of 1.5. Figures 1a through 1d show the beam characteristics for four different times, 0, 0.25, 0.50, and 0.75, measured in beam-plasma periods. The beam-plasma period for a uniform beam of density n_0 is defined in the usual way as $T_p = 2\pi/\omega_p$, and $\omega_p^2 = q^2 n_0/\varepsilon_0 m$ is the beam-plasma frequency. The phase-space plots show density (plasma) oscillations that are excited by the unbalanced external focusing and internal space-charge forces. The total force alternates at the beam-plasma frequency between focusing and defocusing. The charge distribution always remains uniform so that only linear

32   Emittance Growth from Space-Charge Forces

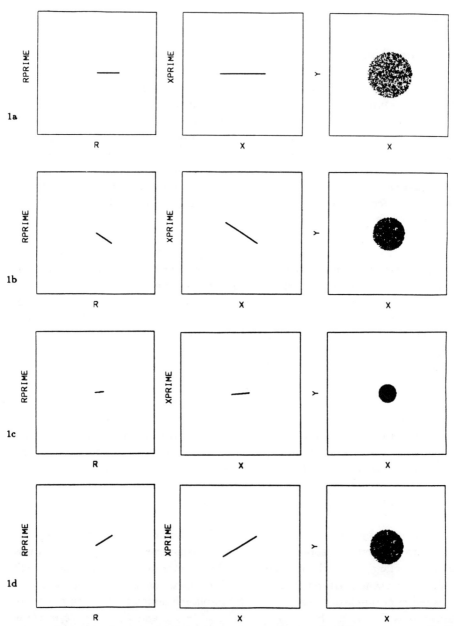

Fig. 1. Radial ($r - r'$) phase space, transverse ($x - x'$ and $y - y'$) phase space, and cross section ($x - y$) from simulation of an initial uniform-density laminar beam in a uniform linear focusing channel for a) $t = 0$, b) $t = 0.25$, c) $t = 0.50$, and d) $t = 0.75$ in units of beam-plasma periods. The initial rms beam sizes in x and y are 50% larger than the matched size.

forces act on the beam, and the emittance remains zero. In the absence of space-charge forces particles are focused by the external fields and cross the axis as they execute betatron oscillations. But because of the repulsive space-charge forces, they do not cross the axis.

3.3 Gaussian Density Laminar Beam

Next we examine the dynamics of an initial Gaussian-density laminar beam with zero-initial-velocity spread, which is rms mismatched by the same factor 1.5. Figures 2a through 2l show a sequence of plots for different times in units of the plasma period (defined for the equivalent uniform beam with the same rms size). For this case, the external force is linear, but the space-charge force is nonlinear. Several new features are obvious. Most of the small amplitude trajectories undergo plasma oscillations (they do not cross the axis) and form an inner core. The large amplitude trajectories correspond to betatron oscillations (they cross the axis) and form an outer halo. In $r-r'$ space the halo looks like a ring-shaped filament. In $x-x'$ space the ring projects into a low-density disk. The projection effect is the result of the fact that any arbitrary point in $r-r'$ space projects to a straight line in $x-x'$ space that passes through the origin and ranges between $-r \leq x \leq r$. Although emittance growth has often been identified with a process of filamentation, we see that the filamentary halo in this problem is observed in the $r-r'$ phase space. In the usual $x-x'$ projected phase space this becomes a diffuse disk-like halo.

Even within a few plasma periods the nonlinear space charge force randomizes the distribution of points within the core. This randomization or thermalization is the result of a process in which the inner part of the filament in $r-r'$ space is stretched and folded many times. The stretching and folding is associated with local variations of the magnitude and sign of the space-charge force caused by local density variations.

The halo produced after several plasma periods is a common feature of all the space-charge mechanisms of emittance growth. We find that the outer filaments seen in $r-r'$ space contain mostly the particles with large initial amplitudes but also contain some particles with small initial amplitudes that were launched during the initial stages of randomization of the core. For our example, the halo is a very ordered structure in $r-r'$ space even after 20 plasma periods; unlike the core, the halo is not yet thermalized.

At present there is no established criterion for defining the halo. For the present example of an rms-mismatched Gaussian laminar beam, we find that an ellipse with the same Courant-Snyder parameters as the rms ellipse and with an emittance five times larger than the rms ellipse appears to enclose the core and exclude most of the halo. If we define the core particles to be all those contained within this ellipse, and define halo particles as those outside, we find that after

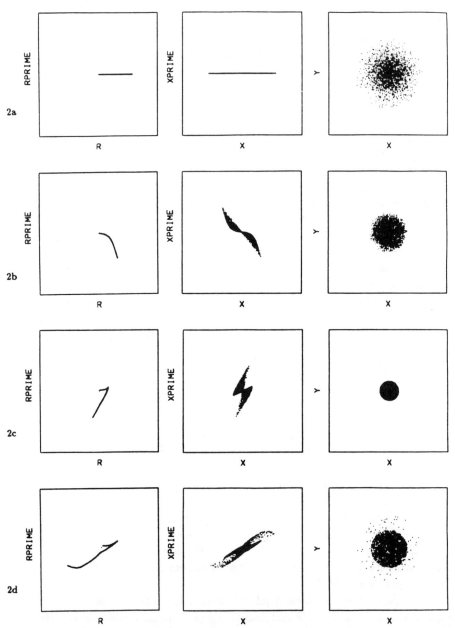

Fig. 2. Radial ($r - r'$) phase space, transverse ($x - x'$ and $y - y'$) phase space, and cross section ($x - y$) from simulation of an initial Gaussian-density laminar beam in a uniform linear focusing channel for a) $t = 0$, b) $t = 0.25$, c) $t = 0.50$, d) $t = 0.75$, e) $t = 1.00$, f) $t = 1.50$, g) $t = 2.00$, h) $t = 3.00$, i) $t = 4.00$, j) $t = 5.00$, k) $t = 10.00$, and l) $t = 20.00$ in units of beam-plasma periods. The initial rms beam sizes in x and y are 50% larger than the matched size.

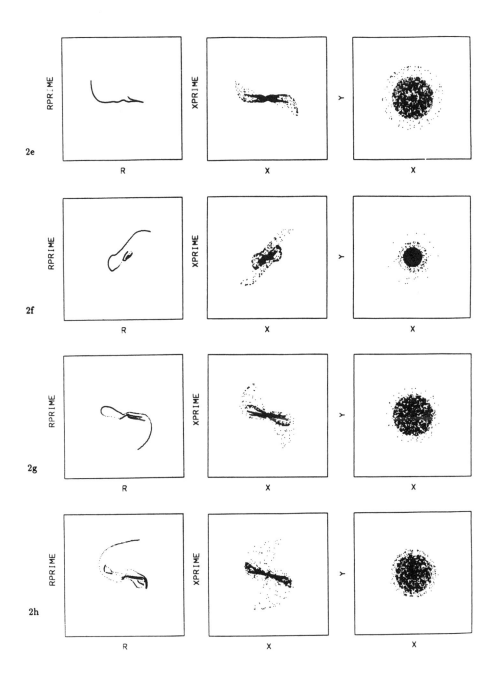

Fig. 2. (cont.)

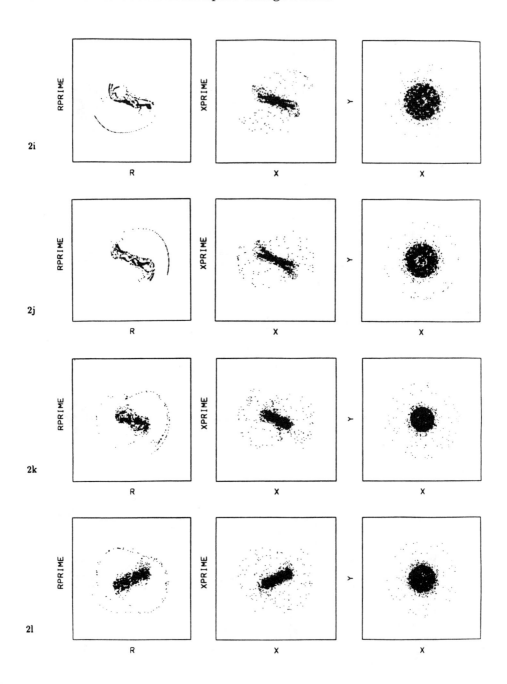

Fig. 2. (cont.)

about 10 plasma periods, 6% of the particles are contained within the halo. For this example the core and the halo contribute about equally to the final rms emittance. Furthermore, the rms emittance of the core grows to its final value in about one-quarter of a plasma period, like that of an rms-matched beam through the charge-redistribution mechanism.[20] The emittance growth of the halo occurs over about 10 plasma periods. We need more studies to determine how these results vary with the amount of mismatch and to determine what happens when using more realistic beams with nonzero initial emittance.

4. CONCLUSIONS

We have discussed the rms-emittance growth caused by nonlinear collective space-charge forces, which is an important cause of loss of brightness for intense low-velocity beams. We have seen that four important emittance-growth mechanisms can be identified on the basis of different sources of free energy available for the growth. Numerical simulation studies have been of great value in giving us a better understanding of these complex phenomena. We have presented an example of the dynamics of a mismatched beam in the extreme space-charge limit, and we have seen that the rms-emittance growth is associated with the formation of both a core and a halo.

Even after more than 20 years we find that there are many questions about space-charge-induced emittance growth that have not been resolved. Many such questions are posed by Lawson[3] in these proceedings. Perhaps an important general question concerns the nature of the state of the beam after a few tens to a few hundred plasma periods, which represent a time scale of practical interest for many linear accelerators and transport systems. Is the beam or at least the core of the beam in some approximate equilibrium state? Is a Maxwell-Boltzmann distribution a good description? If equipartitioning is a characteristic of the beam, why is it so? One interesting and plausible hypothesis is that enough chaos is provided by the nonlinear space-charge forces for the beam to approach a state of maximum entropy. This may explain why beams in numerical simulations do not evolve towards highly ordered equilibrium states like the K-V distribution, and why we observe a tendency for beams to equipartition their kinetic energies. The entropy concept was first applied to beams, and its relationship to rms emittance was explored by Lawson, Lapostolle, and Gluckstern.[38] Recently, the maximum-entropy hypothesis was used to calculate the characteristics of the final distribution for a high-intensity expanding beam in free space.[39] It is clear that a lot of work still remains before we have a complete understanding of this interesting and important area of charged-particle-beam physics.

5. ACKNOWLEDGEMENTS

I would like to thank Martin Reiser for suggesting this topic for the meeting, which was also a celebration of his 60th birthday. I also acknowledge the help of A. Cucchetti on the numerical calculations presented, the helpful suggestions of K. R. Crandall, whose ideas again proved so valuable, and the assistance of Rene Mills who worked with me several years ago attempting to unravel the mysteries of halo formation. Finally, I wish to acknowledge several stimulating discussions with John Lawson, who visited Los Alamos after this meeting, before the written paper was finished.

REFERENCES

1. T. P. Wangler, G. P. Lawrence, T. S. Bhatia, J. H. Billen, K. C. D. Chan, R. W. Garnett, F. W. Guy, D. Liska, S. Nath, G. H. Neuschaefer, and M. Shubaly, "Linear Accelerator for Production of Tritium: Physics Design Challenges," in "Proceedings of the 1990 Linear Accelerator Conference," Los Alamos National Laboratory report LA-12004-C, (September 1990), pp. 548-552.
2. C. Ankenbrandt and S. D. Holmes, "Limits in the Transverse Phase-Space Density in the Fermilab Booster," in "Proceedings of the 1987 IEEE Particle Accelerator Conference," (Washington, D.C., March 16-19, 1987), p. 1066.
3. J. D. Lawson, "The Emittance Concept," these proceedings.
4. I. M. Kapchinskij, *Theory of Resonance Linear Accelerators*, Harwood Academic Publishers (1985), p. 273.
5. I. M. Kapchinskij, and V. V. Vladimirskij, "Limitations of Proton Beam Current in a Strong Focusing Linear Accelerator Associated with the Beam Space Charge, in "Proceedings of the International Conference on High-Energy Accelerators and Instrumentation," (CERN, Geneva, Switzerland 1959), p. 274.
6. Frank J. Sacherer, "RMS Envelope Equations with Space Charge," in *IEEE Trans. Nucl. Sci.* Vol. **18** p. 1105.
7. R. Chasman, "Numerical Calculations of the Effects of Space Charge on Six Dimensional Beam Dynamics in Proton Linear Accelerators," in "Proceedings of the 1968 Proton Linear Accelerator Conference," Brookhaven National Laboratory report BNL 50120 (1968), p. 372.
8. R. Chasman, "Numerical Calculations on Transverse Emittance Growth in Bright Linac Beams," in *IEEE Trans. Nucl. Sci.*, (1969), Vol. NS-16, p. 202.
9. John W. Staples and Robert A. Jameson, "Possible Lower Limit to Linac Emittance," in *IEEE Trans. Nucl. Sci.*, (1979), Vol. NS-26, (3), p. 3698.
10. P. M. Lapostolle, "Possible Emittance Increase Through Filamentation Due to Space Charge in Continuous Beams," in *IEEE Trans. Nucl. Sci.*, (1971), Vol. 18 (3), p. 1101.
11. P. M. Lapostolle, "Round Table Discussion of Space Charge and Related Effects," in "Proceedings of the 1968 Proton Linear Accelerator Conference," Brookhaven National Laboratory report BNL 50120 (1968), p. 437.
12. I. Hofmann and I. Bozsik, "Computer Simulation of Longitudinal-Transverse Space Charge Effects in Bunched Beams," in "Proceedings of the 1981 Linear Accelerator Conference," Los Alamos National Laboratory report LA-9234-C (1982) p. 116.
13. I. Hofmann, "Emittance Growth of Ion Beams with Space Charge," NIM **187**, (1981), p. 281.

14. Robert L. Gluckstern, "Oscillation Modes in Two Dimensional Beams," in "Proceedings of the 1970 Linear Accelerator Conference," Fermi National Accelerator Laboratory, (September 1970), p. 811.
15. I. Hofmann, L. J. Laslett, L. Smith, and I. Haber, "Stability of the Kapchinskij-Vladimirsky (K-V) Distributions in Long Periodic Transport Systems," *Particle Accelerators* **13**, 145 (1983).
16. R. A. Jameson, "Equipartitioning in Linear Accelerators," in "Proceedings of the 1981 Linear Accelerator Conference," Los Alamos National Laboratory report LA-9234-C (October 1982), p. 125.
17. T. P. Wangler and F. W. Guy, "The Influence of Equipartioning on the Emittance of Intense Charged-Particle Beams," in "Proceedings of the 1986 Linear Accelerator Conference," Stanford Linear Accelerator Center, June 2-6, 1986, SLAC report 303, p. 340.
18. F. W. Guy and T. P. Wangler, "Numerical Studies of Emittance Exchange in 2-D Charged Particle Beams," in "Proceedings of the 1986 Linear Accelerator Conference," Stanford Linear Accelerator Center, June 2-6, 1986, SLAC report 303, p. 340.
19. J. Struckmeier, J. Klabunde, and M. Reiser, *Particle Accelerators*, **15**, 47 (1984).
20. T. P. Wangler, K. R. Crandall, R. S. Mills, and M. Reiser, "Relationship Between Field Energy and rms Emittance in Intense Particle Beams," in *IEEE Trans. Nucl. Sci.* **32**, 2196 (1985).
21. T. P. Wangler, K. R. Crandall, and R. S. Mills, "Emittance Growth from Charge Density Changes in High-Current Beams," in "Proceedings of the International Symposium on Heavy Ion Fusion," (Washington, D. C., 1986) AIP Conf. Proc. 152 (1986), p. 166.
22. O. A. Anderson, "Internal Dynamics and Emittance Growth in Space-Charge-Dominated Beams," *Particle Accelerators* **21**, 197 (1987).
23. T. P. Wangler, R. S. Mills, and K. R. Crandall, "Emittance Growth in Intense Beams," in "Proceedings of the 1987 Particle Accelerator Conference," IEEE Catalog No. 87CH2387-9 (1987), p. 1006.
24. M. Reiser, "Free Energy and Emittance Growth in Nonstationary Charged Particle Beams," in "Proceedings of the 1991 Particle Accelerator Conference," (San Francisco, California, May 1991).
25. A. Cucchetti, M. Reiser, and T. P. Wangler, "Simulation Studies of Emittance Growth in RMS Mismatched Beams," in "Proceedings of 1991 Particle Accelerator Conference," (San Francisco, California, May 1991).
26. I. Hofmann and J. Struckmeier, "Generalized Three-Dimensional Equations for the Emittance and Field Energy of High-Current Beams in Periodic Focusing Structures," *Particle Accelerators* **21**, 69 (1987).
27. I. Hofmann, "Space-Charge Dominated Beams," US-CERN Joint Topical Course, "Frontiers of Particle Beams," (South Padre Island, Texas, October 23-29, 1986), GSI Report GSI-87-40, (July, 1987).
28. J. Struckmeier and M. Reiser, "Theoretical Studies of Envelope Oscillations and Instabilities of Mismatched Intense Charged Particle Beams in Periodic Focusing Channels and Particle Accelerators," *Particle Accelerators* **14**, 227 (1984).
29. M. E. Jones and B. E. Carlsten, "Space-Charge-Induced Emittance Growth in the Transport of High-Brightness Electron Beams," in "Proceedings of the 1987 IEEE Particle Accelerator Conference," IEEE Catalog No. 87CH2387-9, (1987), p. 13.
30. B. E. Carlsten and R. L. Sheffield, "Photoelectric Injector Design Considerations," in "Proceedings of the 1988 Linear Accelerator Conference," (Williamsburg, Virginia, October 1988), p. 1319.
31. R. L. Sheffield, "Photocathode Electron Guns," in Physics of Particle Accelerators, AIP Conf. Proceedings **184**, Vol. 2, p. 1500.
32. B. E. Carlsten, "New Photoelectric Injector Design for the Los Alamos National Laboratory XUV FEL Accelerator," *Nucl. Inst. and Methods in Phys. Research* **A285**, 313-319 (1989).

33. H. Hanerfeld, W. Herrmannsfeldt, and R. H. Miller, "Higher Order Correlations in Computed Particle Distributions," in "Proceedings of the 1989 Particle Accelerator Conference," IEEE Catalog Number 89CH2669-0, (1989), p. 880.
34. I. Haber, "High-Current Simulation Codes, in the High Current, High Brightness, High Duty Factor Ion Injectors," AIP Conference Proceedings **139**, 107 (1985).
35. I. Haber and H. Rudd, "Numerical Limits on P.I.C. Simulation of Low Emittance Transport," in Linear Accelerator and Beam Optics Codes, AIP Conference Proceedings **177**, (1988) p. 161.
36. F. W. Guy, P. M. Lapostolle, and T. P. Wangler, "The Influence of Density Distribution on the Stability of Beams," in "Proceedings of the 1987 Particle Accelerator Conference," (Washington, D.C., March 16-17), 1987, IEEE Catalog No. 87CH2387-9, p. 1149.
37. K. R. Crandall, R. S. Mills, and T. P. Wangler, "Simulation of Continuous Beams Having Azimuthal Symmetry to Check the Relation Between Emittance Growth and Nonlinear Energy," Los Alamos National Laboratory Group AT-1 memorandum AT-1:85-216, June 12, 1985.
38. J. D. Lawson, P. M. Lapostolle, and R. L. Gluckstern, "Emittance, Entropy, and Information," *Particle Accelerators* **5**, 61 (1973).
39. J. S. O'Connell, "Limiting Density Distribution for Charged Particle Beams in Free Space," presented at 1981 Particle Accelerator Conference, (San Francisco, California, May 6 to 9, 1991), (to be published).

ADVANCES IN THE THEORY OF CHARGED PARTICLE BEAM TRANSPORT

Alex Dragt
Department of Physics and Astronomy
University of Maryland, College Park, MD 20742 USA

ABSTRACT

A brief overview will be given of recent advances in the theoretical treatment of charged particle beam transport. Topics covered will include nonlinear aberration compensation, the generalization of emittance concepts to four and six dimensional phase spaces, and the calculation and correction of aberrations arising from linear and nonlinear space charge forces.

SOLITONS AND PARTICLE BEAMS*

J. J. Bisognano
Continuous Electron Beam Accelerator Facility,
12000 Jefferson Avenue, Newport News, VA 23606

ABSTRACT

Since space charge waves on a particle beam exhibit both dispersive and nonlinear character, soliton-like behavior is possible. Some theoretical aspects of dispersive, nonlinear wave propagation in high brightness beams are discussed. Numerical examples for realizable beams are presented, and issues for future studies are noted.

INTRODUCTION

Space charge forces can produce longitudinal density waves in low momentum spread, charged particle beams.[1] For a uniform beam of radius a transported in a perfectly conducting beampipe of radius b, the propagation is nondispersive in the linear, long wavelength approximation. The wave velocity v_p is

$$v_p = \frac{\omega}{k} = \sqrt{\frac{e^2 \lambda_0 g}{4\pi\epsilon_0 m}} \qquad (1)$$

where ω is the mode frequency for wave number k, e is the electron charge, λ_0 is the unperturbed linear particle density, $g = 1 + 2 \log b/a$, ϵ_0 is the permittivity of free space, and m is the mass of the beam particles. However, for large density perturbations nonlinearity cannot be ignored, and for short wavelengths (small compared to the beampipe dimension) the propagation is dispersive with the wave velocity dependent on wavelength. For many physical systems[2] this combination of nonlinearity and dispersion leads to solitary waves and solitons. In fact, this is the case for the illustrative particle beam configuration discussed in this paper.

SOLITARY WAVES AND SOLITONS

Nonlinearity in wave propagation typically leads to steepening phenomena. For example, consider the simple[3] wave equation

$$u_t + (1+u)u_x = 0 \qquad (2)$$

---

* Supported by D.O.E. contract #DE-AC05-84ER40150

which has the implicit solution

$$u(x,t) = f(x - (1+u)t) \quad (3)$$

where f is an arbitrary differentiable function. Note the velocity, $(1+u)$, depends on the amplitude, and, in particular, higher amplitudes propagate faster. If f describes a localized distribution, the peak value will tend to overtake lower values, and steepening and breaking of the pulse will result. On the other hand, if the velocity depends strongly on wavelength (dispersion), a localized distribution spreads as it propagates. A solitary wave results when the nonlinear steepening is canceled by the dispersive spreading, yielding a localized disturbance which propagates without distortion. Since solitary waves of different heights will generally travel with different velocities, collisions can occur. The term soliton describes solitary waves which maintain their identity and shape after collision.

SPACE CHARGE FORCES

For a beam in a beampipe, the longitudinal force F generated by longitudinal density variations is described by

$$F = \frac{-ge^2}{4\pi\epsilon_0} \frac{\partial \lambda}{\partial z} \quad (4)$$

in the long wavelength limit for density λ. In k-space, the spatial Fourier transform $\tilde{F} \propto ik\tilde{\lambda}$. More generally, the Green's function for a cylindrically symmetric distribution in a cylindrical symmetric pipe is

$$G(\rho, z; \rho', z') = \frac{1}{4\pi\epsilon_0} \frac{2}{\pi b^2} \int_{-\infty}^{\infty} dk \sum_{n=1}^{\infty} (-ik) e^{ik(z-z')} \frac{J_0(\frac{x_n\rho}{b})J_0(\frac{x_n\rho'}{b})}{((\frac{x_n}{b})^2 + k^2)J_1^2(x_n)} \quad (5)$$

where x_n is the nth zero of the Bessel function J_0. Note that for small $k \ll (x_n/b)$, ik behavior dominates.

Consider a distribution of the form $J_0(x_1\rho/b)e^{ikz}$. In a linearized fluid model, this function describes a perturbation eigenmode of a uniform beam filling the beampipe. The underlying force law is modified from

$$ik \longrightarrow \frac{ik}{1+\alpha k^2} \quad (6)$$

where $\alpha = b^2/x_1^2$. The phase velocity

$$v_{\text{phase}} = \frac{v_p}{\sqrt{1+\alpha k^2}} \quad (7)$$

where the g implicit in v_p is now a geometric factor of order unity, and the propagation has become dispersive. On expanding the denominator of the right side of relation (6) for small α, we note that a third derivative term $(-ik^3)$ is added to the first derivative term (ik). This is suggestive of the structure of the Kortweg-DeVries (KdV) equation, which exhibits soliton behavior.

1-D NONLINEAR FLUID MODEL

As a first step in understanding the interplay of nonlinearity and dispersion for space charge dominated beams, we analyze a 1-D nonlinear cold fluid model of a uniform beam with the force law given in relation (6). Admittedly, some possibly important transverse effects may be lost. With $v_p = 1$, the fluid equations are

$$\frac{\partial n}{\partial t} + \frac{\partial}{\partial x}(nv) = 0 \tag{8}$$

$$\frac{\partial v}{\partial t} + v\frac{\partial v}{\partial x} = -\frac{\partial \Phi}{\partial x} \tag{9}$$

$$n(x,t) = n_0 + n_1(x,t) \tag{10}$$

$$\tilde{\Phi}(k) = \frac{\tilde{n}_1}{1 + \alpha k^2}. \tag{11}$$

At this point we can parallel Davidson's discussion of ion-acoustic solitary waves,[4] and look for solutions of the form $n_1(qx - \omega t)$, $v(qx - \omega t)$, etc. which roll-off at $\pm\infty$. Equations (8)-(11) imply that

$$n = \frac{n_0}{1 - \frac{q}{\omega}v} \tag{12}$$

$$\left(\frac{\omega}{q}\right)^2 = \left(\frac{\omega}{q} - v\right)^2 - 2\Phi \tag{13}$$

and for localized Φ

$$\frac{\alpha q^2 \Phi'^2}{2} - \frac{\Phi^2}{2} - n_0 \left(\frac{\omega}{q}\right)^2 \sqrt{1 - 2\left(\frac{q}{\omega}\right)^2 \Phi} = 0 \tag{14}$$

where $'$ denotes differentiation. The resulting first order equation (14) is easily solved numerically for Φ, n, and v to yield the pulse shape of the self-consistent solitary waves as a function of the parameter ω/q. The peak value of Φ is given by

$$\Phi_{\text{peak}} = 2\left(\frac{\omega}{q} - 1\right) \tag{15}$$

and the peak density is given by

$$n_{\text{peak}} = \frac{n_0}{1 - 2\left(\frac{q}{\omega}\right)^2 \Phi}. \tag{16}$$

When $\omega/q = 2$, $\Phi = 2$, and the density n becomes singular, indicating breaking.

A multiple time scale analysis of these fluid equations with (ω/q) as the small expansion parameter yields the KdV equation as the lowest approximation. The KdV soliton, however, does not exhibit breaking. This difference for large (ω/q) is traceable to the weaker high frequency dispersion associated with the

$$\frac{k}{1 + \alpha k^2} \tag{17}$$

behavior of the space charge force versus the

$$k - \alpha k^3 \tag{18}$$

behavior implicit in the KdV equation.

CONCLUSIONS

A simple, 1-D model of longitudinal space charge waves exhibits solitary waves together with breaking at large amplitudes. Clearly, this analysis represents only a first step in understanding, and many questions remain open. Of most importance are the complications introduced by the transverse distribution and betatron oscillations. Although $J_0(x_1\rho/b)e^{ikz}$ provides a self-consistent mode for the linearized equations, this transverse distribution is not self-consistent for the nonlinear system. The full Green's function, with the infinite sum exhibited in equation (5), needs to be addressed. Also, the assumption of transport of a high current beam of the same dimension as the beampipe simplified the mathematics (collapsing the infinite sum), but it is not practical experimentally. Wall resistance and the associated slow growing instability would complete the picture.

Whether these solitary waves are indeed solitons is not clear, even in the 1-D model presented. Whitham[4] has studied a similar force law in a model of water waves and found preservation of wave shape after the collision of two such localized pulses. He also found some interesting phenomena associated with breaking. Both one and two dimensional simulations would be valuable in investigating these issues more thoroughly.

The scaling of possible experiments is set by the parameter v_p given in equation (1). For example, breaking occurs when $\omega/q = 2$ in units of v_p, and the solitary wave velocity lies between v_p and $2v_p$. Low energy ($\beta = 0.3$) electron beams[5] found in high-space-charge transport experiments can take values of v_p approaching 10^7 m/s. Since dispersive effects are expected for pulse lengths of the order of the beampipe radius, typically centimeters, it appears that several meters of transport may be sufficient to observe some of the phenomena discussed. Ion storage rings may also offer some possibilities, although the microwave instability could be a problem.

ACKNOWLEDGMENT

This work derives from a collaboration with Kurt Riedel some years ago when he was a summer college intern at Lawrence Berkeley Laboratory.

REFERENCES

1. J. Bisognano, I. Haber, L. Smith, and A. Sternlieb, *IEEE Transactions in Nuclear Science*, **NS-28**, 2513 (1983).

2. R. K. Dodd, J. C. Eilbeck, J. D. Gibbon, and H. C. Morris, *Solitons and Nonlinear Wave Equations*, Academic Press, Orlando (1982).

3. P. G. Drazin and R. S. Johnson, *Solitons: an Introduction*, Cambridge University Press, Cambridge (1989).

4. R. C. Davidson, *Methods in Nonlinear Plasma Theory*, Academic Press, New York (1972).

5. B. Fornberg and G. B. Whitham, *Proc. R. Soc. Lond.*, **289**, 373(1978).

6. T. Shea, et al., *Proc. of the 1989 Particle Accelerator Conference*, IEEE 89CH2669-0, 1049 (1989).

EXPERIMENTAL STUDIES OF EMITTANCE GROWTH IN A NONUNIFORM, MISMATCHED, AND MISALIGNED SPACE-CHARGE DOMINATED BEAM IN A SOLENOID CHANNEL[*]

D. Kehne, M. Reiser
Laboratory for Plasma Research
University of Maryland, College Park, Maryland 20742

H. Rudd
Rudd Consulting Services
6922 Nashville Rd., Lanham, MD 20706

ABSTRACT

Experimental and numerical studies of emittance growth resulting from beam mismatch have been performed at the University of Maryland Electron Beam Transport Experiment. A 5-beamlet distribution of energy 5 keV and total current 44 mA passes through two-solenoid matching lenses and into a 36-solenoid transport channel. Theory predicts substantial emittance growth due to density nonuniformity, beam mismatch, and beam misalignment. Experimentally, the 5-beamlet configuration is mismatched and transported through the channel. Random misalignments in the channel produce a gradually growing offset in the beam. The final emittance is measured for the mismatched beam. Simulation results of the matched and mismatched beams reveal that a large halo accounts for the emittance growth in the mismatched case. This is supported by experiment. The difference between the experimental emittance data and the numerical simulation is analyzed and attributed to the fact that the density of the halo is below the detection threshold of the emittance meter. Simulation also predicts that misalignments contribute negligibly to the emittance growth. Excellent agreement is found between fluorescent screen images of beam structure and simulation results.

INTRODUCTION AND THEORY

Past theoretical, experimental, and numerical studies [1-6] have shown that the free energy associated with nonuniform charge distributions is converted to random kinetic energy resulting in emittance growth. In addition, mismatched [3,6,7] and misaligned beams are similarly characterized by excess free energy and also undergo emittance growth. Recent formulation [6] of the theory offers a quantitative prediction of the emittance growth due to space charge homogenization, beam mismatch, and beam misalignment. In conjunction with numerical analysis, this experiment attempts to study the effects of these theoretically predicted emittance growth mechanisms.

The formulation of the theory reviewed here [6] incorporates the associated change in beam radius with the emittance growth resulting from the conversion of free energy in nonuniform, mismatched, and misaligned beams.

The predicted emittance growth is given by the equation

---

[*] Research supported by the U.S. Department of Energy

$$\frac{\varepsilon_f}{\varepsilon_i} = \frac{a_f}{a_i}\left[1 + \frac{\sigma_0^2}{\sigma_i^2}\left(\frac{a_f^2}{a_i^2} - 1\right)\right]^{1/2}, \qquad (1)$$

where σ_0 is the particle phase advance per period neglecting space charge, σ_i is the initial phase advance per period in the presence of space charge, a_i is the effective (2×RMS) radius of the equivalent initial stationary beam, and a_f is the effective radius of the equivalent final stationary beam. For period length S, the value of σ_i can be calculated using the equation [8]

$$\frac{\sigma_i^2}{S^2} = \frac{\sigma_0^2}{S^2} - \frac{K}{a_i^2}, \qquad (2)$$

where $K = 2(I/I_0)/(\beta\gamma)^3$ is the generalized perveance, $I_0 = 1.7 \times 10^4$ A is the characteristic current for electrons, I is the total beam current, $\beta = v/c$, v is the particle velocity, and the relativistic energy factor γ is given by $(1-\beta^2)^{-1/2}$. The ratio a_f/a_i required in Eq. (1) is calculated from

$$\left(\frac{a_f}{a_i}\right)^2 - 1 - \left(1 - \frac{\sigma_i^2}{\sigma_0^2}\right)\ln\frac{a_f}{a_i} = h, \qquad (3)$$

where h is the free energy parameter.

For nonuniform charge distributions, h is defined as

$$h = h_s = \frac{1}{4}\left(1 - \frac{\sigma_i^2}{\sigma_0^2}\right)\frac{U}{w_o}, \qquad (4)$$

where U/w_o is a dimensionless parameter depending only on the geometry of the nonuniform distribution.

For mismatched beams, $h = h_m$ is defined as

$$h = h_m = \frac{1}{2}\frac{\sigma_i^2}{\sigma_0^2}\left(\frac{a_i^2}{a_0^2} - 1\right) - \frac{1}{2}\left(1 - \frac{a_0^2}{a_i^2}\right) + \left(1 - \frac{\sigma_i^2}{\sigma_0^2}\right)\ln\frac{a_i}{a_0}, \qquad (5)$$

where a_0 is the effective radius of the mismatched beam waist at the beginning of the channel.

For misaligned beams, $h = h_c$ is defined as

$$h = h_c = \left(\frac{x_c}{a_i}\right)^2 \frac{\sigma_c^2}{\sigma_0^2}, \tag{6}$$

where x_c is the initial offset of the beam and $\sigma_c^2 = \sigma_0^2 - \sigma_{im}^2$. Here, $\sigma_{im}^2 = K/b^2$ is the image force factor and b is the beam pipe radius.

In the presence of all three effects, the associated free energy parameters h_s, h_m, and h_c are first added and then the radius growth and emittance growth calculated.

EXPERIMENTAL SETUP

The apparatus used to test this theory is the high perveance thermionic electron gun and transport channel of the University of Maryland Electron Beam Transport Experiment [9]. The channel consists of 36 periodically spaced solenoids (period S = 13.6 cm). Two solenoids identical to the channel solenoids are used to match the beam to the channel. As in previous work with multiple beams [9], the full 5 kV, 240 mA solid beam (2 μsec pulse length, 60 Hz repetition rate) produced by the gun is apertured to form a 5-beamlet configuration of 44 mA total current. The aperture creates a beam such that one beamlet is located on-axis and the other four are placed symmetrically around the center one. The radius of each beamlet is 1.19 mm and the distance between the center of the inner beamlet and the center of each outer beamlet is 3.57 mm. The calculated [5] value of the geometrical factor U/w_0 in Eq. (4) is 0.2656. The initial effective radius and the unnormalized effective (4×RMS) initial emittance of the 5-beamlet structure were calculated [5] to be 4.67 mm and 64.8 mm-mrad, respectively.

The changing structure of the beam can be observed at any point in the channel or matching section using a phosphor screen mounted on an axially translatable trolley. Furthermore, the beam emittance can be measured at the end of the channel using a slit/pinhole apparatus detailed elsewhere [9]. The slit apertures sheet beamlets at different locations across the beam. After being allowed to expand, the profile across each sheet beam is measured and fit to a Maxwellian distribution. Further analysis generates the RMS emittance. This type of analysis assumes that the beam is axisymmetric. This assumption is only valid if the 5-beamlet configuration has merged sufficiently to lose its nonaxisymmetric structure.

As in past studies, the Particle-In-Cell code SHIFTXY has been used to simulate the beam. A thorough description of the code can be found elsewhere [9] and hence will be deferred here. The model used in the simulations for the on-axis magnetic field is an analytically fit field given by

$$B(z) = B_0 \frac{\exp(-z^2/2b^2)}{1 + z^2/a^2}, \tag{7}$$

where B_0 is the peak magnetic field, a = 4.40 cm, and b = 2.29 cm. Off-axis fields are derived by Taylor expansion of this equation. Using this analytic representation of the on-axis magnetic focusing field, the phase advance per period neglecting space charge was calculated to be 77°.

The research discussed in this paper focuses on a beam that was drastically RMS-mismatched in order to trigger envelope oscillations. For comparison purposes, some data involving the RMS-matched beam is also presented.

MISMATCHED BEAM RESULTS

Parameters for the matched beam case, that have been presented elsewhere [9], are repeated here for convenience. For the matched beam, theory predicts an emittance growth of 1.52 and a value of a_f/a_i of 1.05. Matched beam simulations showed an emittance growth of about 1.5 and a corresponding growth in average radius of 1.05. This growth occurred within the matching section ($z < 30$ cm from the aperture plate) in accordance with theory [9].

For the mismatched beam, we chose a mismatch ratio of $a_0/a_i = 0.5$. To maintain consistency with the theory, the ratio a_0/a_i is defined as the ratio of mismatched to matched beam radius one half period from the center of the first channel lens. In addition, the beam waist is located at this point. The SHIFTXY code was used to find the matching lens settings that result in a mismatch ratio of 0.5. It was found that these lens settings could also be accurately found assuming an equivalent uniform beam in a K-V envelope code. Since the beam is space-charge dominated, the growth in emittance in the matching section has negligible effect on the envelope.

For the mismatch ratio of $a_0/a_i = 0.5$, equations (1) through (4) predict a total emittance growth of 3.75 and total growth in average radius of 1.30. These numbers are supported by simulation [10]. The simulation results shown in Fig. 1 and Fig. 2 are the growth in effective radius and the growth in effective emittance plotted as a function of period number N. The value N=0 corresponds to the waist located one-half period before the first channel lens, the value of N=1 corresponds to the waist 1 period after N=0, N=2 corresponds to the waist 2 periods after N=0 and so forth. In Fig. 1, the beam, though drastically mismatched at the beginning of the channel is nearly matched at the end, with only small envelope oscillations still evident. Fig. 2 reveals the total emittance growth occurring entirely within 15 periods, less than half of the total channel length.

Similar emittance results were expected experimentally at the end of the channel. Beam pictures taken at many locations along the channel verified that the simulation was a fairly accurate representation of the actual experiment. Some of the pictures of the mismatched beam, located 30.6cm (N = 0), 44.2 cm (N = 1), 85.0 cm (N=4), 193.8 cm (N = 12), and 524.2 cm (N = 36) from the aperture plate are shown in Fig. 3 with the corresponding simulation pictures.

In these pictures, the beam rotates counter clockwise. Though the initial angular positions are different by approximately 7°, the beam in the experiment rotates 4% ± 1% faster than the simulated beam. In addition, slight differences with respect to structure become more pronounced as the beam propagates further down the channel. The rotation about the channel axis of a particle with mass m, charge q, and canonical angular momentum p_θ in a magnetic field B(z) can be calculated using the well-known equation

$$\theta - \theta_0 = \int_{z_0}^{z} \left(\frac{-qB(z)}{2\gamma m \beta c} + \frac{p_\theta}{\gamma m \beta c r^2} \right) dz, \tag{8}$$

where β and γ are as defined before, c is the speed of light, and $\theta-\theta_0$ is the amount of rotation about the axis between axial locations z_0 and z. Fig. 4 shows a comparison of the measured on-axis field and the analytic field defined by Eq. (7). By analyzing each of the fields in Fig. 4, it was discovered that the integral of the analytic field over one period is 5.5% ± 1% less than the same integral of the measured field. The inaccuracy in the tail of the analytic field is the source of the discrepancy in beam rotation found between experiment and simulation.

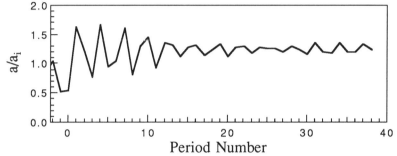

Fig. 1. Ratio of the effective radius a to initial effective radius of the beam a_i plotted versus period number for the mismatched beam.

When the mismatched beam emittance was measured at the end of the channel (z = 524 cm) [10], a value of 123 mm-mrad ($\varepsilon_f/\varepsilon_i \approx 1.9$) resulted. Due to substantial noise, the error of this measurement is about 20%. Even so, the value for the measured emittance of the mismatched beam is less than half that predicted by both theory (242 mm-mrad) and simulation (265 mm-mrad).

The picture at z = 524 cm, taken at the end of the channel, shows that the beam is round and symmetric, suitable for emittance measurement with a slit/pinhole system that requires axial symmetry. Unfortunately, the CCD camera used to take these pictures did not record with adequate sensitivity the very diffuse halo surrounding the mismatched beam which is visible to the eye on the fluorescent screen. The first indication of a halo occurs after about 5 channel periods. After 6 periods, the inner core retains substantial azimuthal structure and the halo is distinct. After 12 periods (z = 193.8 cm), the emittance growth is complete, the structure of the inner core has nearly disappeared, and the halo extends to twice the radius of the core.

Due to the low particle density, the outer portion of the beam comprising the halo was also below the sensitivity threshold of the emittance meter located at the end of the channel. In order to see the effect of the halo on emittance, the simulation was used to remove halo particles from the emittance calculation For comparison, this was done for both the matched and mismatched beams. The elimination of the halo was done in two steps. First, the real-space (x-y) halo was removed. This revealed the emittance as a function of beam radius and intensity presented in Table I. The mismatched beam results imply that particles located from 6.75 to 12 mm from the beam center, comprising 10% of the beam current, are responsible for about 50% of the increase in emittance over that of the matched beam. The matched beam shows no evidence of a halo. The emittance meter measured the mismatched beam emittance within radius 6.75 mm and the matched beam emittance within 5.25 mm. Therefore, the value of emittance growth in Table I that corresponds to the measured emittance of the mismatched beam is 2.75. This corresponds to a final emittance of 178 mm-mrad. This is 45% higher than the measured value of 123 mm-mrad. This discrepancy is discussed later.

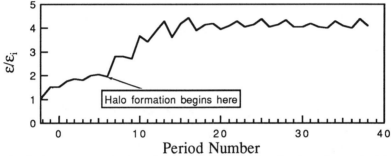

Fig. 2. Emittance growth plotted versus period number for the mismatched beam.

The second step in determining the halo effect involved only the mismatched beam. Here, all halo particles were removed from the emittance calculation. This is necessary because the real space (x-y) halo constitutes only part of the total halo. Some of the high velocity halo particles are crossing the beam axis and hence are included in the emittance calculation in the simulations described in Table I. Data analysis was performed to eliminate all halo particles existing in both real space (x-y) and velocity space (x'-y'). The x-y and x'-y' phase space point plots are shown in Fig. 5. The halo is evident in both cases. All particles with radius greater than 6 mm or with x' or y' greater than half the maximum values of x' and y' were excluded from the emittance calculation. The results revealed that the full halo constitutes approximately 18% of the total beam. Furthermore, the emittance of the particles not in the halo was found to be the same as that for the matched beam (i.e. $\varepsilon_f/\varepsilon_i = 1.53$). This is in agreement with similar work done by Anderson [3]. The low value of the mismatched beam emittance measured in the experiment is primarily the result of excluding part of this halo.

As mentioned earlier, the mismatched beam emittance from the simulation is still 45% higher than the corresponding measured emittance. One possible explanation is that the simulation emittance calculation and experimental measurement were not performed at exactly corresponding locations in the channel. The experimental measurement was performed 12.3 cm past the center of the final channel lens The simulation calculations were performed 6.8 cm past the center of the final channel lens. Since the beam expands over this drift space, the simulation and experimental data do not have one-to-one correspondence with regard to what particles are at what radius. Since the temperature of a beam is larger when the beam is small, in keeping with the conservation of emittance, the inner particles of the simulated beam would have higher temperature than the measured beam. Hence, the values in Table I would be higher than that measured in the experiment. In addition, recall that the experimental magnetic fields are approximately 5% higher than the field used in the simulation. Since this variation in field will cause differences in mismatch envelope oscillations, again different beam sizes could result causing the same discrepancy in emittance values. A careful analysis of this problem is underway in order to improve future emittance measurements.

One interesting feature of the emittance measurements involves the profile shapes of the sampled sheet beamlets. Though there is deviation from the best Maxwellian fit in all the profiles, the mismatched beam profiles fit to Maxwellian distributions noticeably better than the matched beam profiles. This seems to indicate that the mismatched beam is thermalizing at a faster rate than the matched beam. Since the mismatched beam contains the additional mismatch free energy that is not present in the matched beam, this is a reasonable result. The subject of beam-core thermalization of the matched and mismatched distributions is under investigation.

Recall that the theory predicts that misalignment of the beam can also cause emittance growth. In the experimental channel, misalignments are definitely present. Due to random offsets and tilts in the focusing magnets, the beam alignment in the channel increases from less than 0.25 mm off-axis at the aperture plate to 6 mm off-axis at the end of the channel. Though Eq. (6) only predicts emittance growth due to initial offset of the beam, an estimate of misalignment effects can be obtained by calculating the initial offset that would, assuming full emittance growth takes place, result in the emittance that was measured in the experiment. The value of this offset is about 1.2 mm. This falls well within the value of random misalignments. In order to determine the growth rate of emittance due to beam misalignment, simulation of the matched 5-beams with an initial offset of 3.0 mm was performed. For the aligned, matched beam initially offset by 3.0 mm, theory predicts a total emittance growth of 3.8. However, simulations show no misalignment induced emittance growth even after 110 periods. It seems that the beam traverses the entire channel before any emittance growth associated with misalignment can occur. Thus, we conclude that misalignments are not the cause of any measurable emittance growth in this experiment.

Table I. Simulation results of the partial beam emittance and % of total beam current as a function of radius are shown for the matched and mismatched beams. Total growth in the matched beam is 1.53. Initial emittance is 64.8 mm-mrad.

Radius [mm]	% of total current within given radius of matched and mismatched beams		Emittance growth of matched and mismatched beams calculated from particles inside radius	
	Matched Beam	Mismatched Beam	Matched Beam	Mismatched Beam
12	100	100	1.53	4.07
11	100	99.3	1.53	3.97
9	100	95.6	1.53	3.42
8	100	93.4	1.53	3.12
7	100	90.8	1.53	2.81
6.75	100	90.2	1.53	2.75
5.25	99	87.3	1.53	2.41

54 Experimental Studies of Emittance Growth

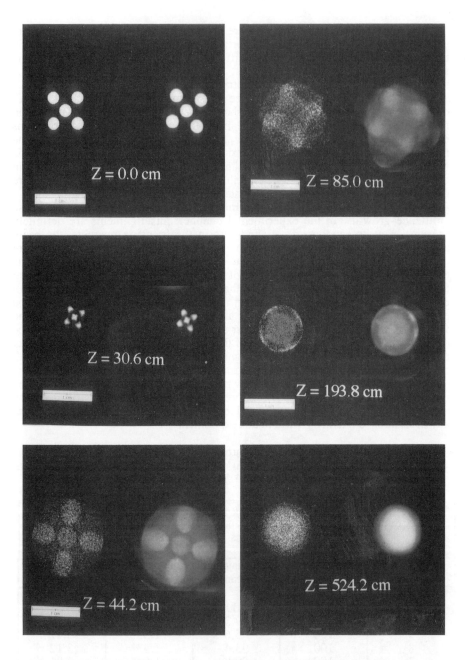

Fig. 3. Each picture above shows a simulation real space plot (left image) and an experimental picture (right image) of the mismatched beam at location z where z is the distance from the aperture plate. The experimental picture at z = 0 cm is the aperture plate backlit by the cathode. The scale in each photo is the same.

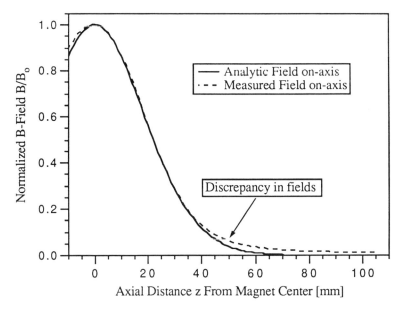

Fig. 4. The on-axis measured and analytic fields are plotted to reveal the small difference in the tails that is responsible for the discrepancy in beam rotation found between experiment and simulation.

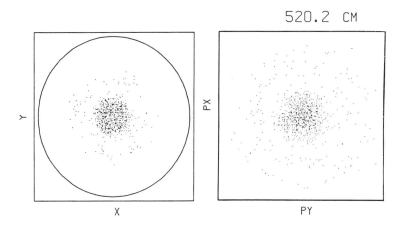

Fig. 5. Simulation x-y and x'-y' phase space point plots for the mismatched beam at z = 520 cm showing the halo in real space and velocity space.

CONCLUSION

Theoretically predicted emittance growth due to the conversion of free energy to random kinetic energy in a nonuniform, mismatched beam has been supported experimentally. The theoretically predicted growth in radius and emittance has been confirmed by simulation. Experiment and simulation both show a strong halo at the channel end. Simulation and emittance measurements of the mismatched beam verify that the real space halo, comprising 10% of the beam, is responsible for about 70% of the emittance growth resulting from the mismatch. Simulation shows that, when the full halo, which includes the velocity space halo as well as the real space halo, is removed from the emittance calculation, there is no additional emittance growth above that generated by the nonuniform distribution. Though discrepancies in the simulated and experimental magnetic fields have been resolved, discrepancies in the final emittance of the mismatched beam still exist. This large discrepancy is primarily due to exclusion of the halo in the measurement.

ACKNOWLEDGEMENTS

The authors would like to thank M. Russotto and J. Li for the computer and photographic work that was necessary to render the pictures of both simulation and experiment.

REFERENCES

1. J. Struckmeier, J. Klabunde, and M. Reiser, Part. Accel., 14, 227 (1984).
2. T.P Wangler, K.R. Crandall, R.S. Mills, and M. Reiser, IEEE Trans. Nucl. Sci. NS-32, 2196, (1985).
3. O.A. Anderson, Part. Accel. 21, 197, (1987).
4. I. Hofmann and J. Struckmeier, Part. Accel., 21, 69, (1987).
5. M. Reiser, C.R. Chang, D.Kehne, K. Low, T. Shea, H. Rudd, and I. Haber, Phys. Rev. Lett., 61, , 2933, (1988).
6. M. Reiser, J. Appl. Phys. **70** (4), 1919 (1991).
7. Tiefenback, M. G., Proceedings of the 1987 Particle Accelerator Conference, March 16-19, Washington, DC, IEEE catalog no. 87CH2387-9, pp. 1046-1048.
8. M. Reiser, Particle Accelerators **8**, 167 (1978).
9. I. Haber, D. Kehne, M. Reiser, and H. Rudd, Phys. Rev. A, **44** (6), 5194 (1991).
10. D. Kehne, M. Reiser, and H. Rudd, presented at the 1991 PAC, May 6-9, San Francisco, to be published IEEE Conference Proceedings of the 1991 Particle Accelerator Conference.
11. M. Reiser, presented at the 1990 HIIF Conference, Monterrey, Ca., Dec. 3-6, 1991; to be published in Particle Accelerators.
12. M.J. Rhee and R. F. Schneider, Particle Accelerators **20**, 133 (1986).
13. D. Kehne and M. Reiser, presented at the 1990 HIIF Conference, Monterrey, Ca., Dec. 3-6, 1991; to be published in Particle Accelerators.

STUDIES OF LONGITUDINAL BEAM COMPRESSION AND RESISTIVE-WALL INSTABILITY*

J. G. Wang, M. Reiser, D. X. Wang, and W. M. Guo
Laboratory for Plasma Research and Department of Electrical Engineering
University of Maryland, College Park, Maryland 20742

ABSTRACT

This paper reports the results of design studies of longitudinal beam compression and resistive-wall instability, and the current status of the two experiments at the University of Maryland.

INTRODUCTION

The longitudinal beam compression and resistive-wall instability experiments are on-going research projects at the University of Maryland electron beam transport facility [1]. The beam physics issues motivating these experiments are important for advanced accelerator applications such as the use of induction linacs as Heavy Ion Inertial Fusion drivers. This paper first discusses the design study of longitudinal beam compression, which concerns drift bunching of the electron beams with a parabolic density distribution and uniform density distribution in the longitudinal direction. The experimental implementation of the theory is presented. The longitudinal instabilities are theoretically studied for the intense beams in a transport channel with complex wall impedances. The growth rates of the instabilities in different wall structures are compared. The design of the experiment with a pure resistive wall is described. An electron beam injector consisting of a variable-perveance gridded electron gun followed by three matching lenses and one induction acceleration module has been constructed to perform these two experiments. The performance characteristics of the injector are presented.

LONGITUDINAL COMPRESSION

1). Drift bunching of the beam with a parabolic density distribution.

The longitudinal compression of a beam pulse can be achieved by drift bunching of the beam with initial velocity tilt. For a parabolic density distribution in the beam length, the longitudinal field due to the space charge is linear and the distribution function is a constant of motion during drift bunching. This compression process can be described by the longitudinal envelope equation [2,3]

$$Z''_m - \frac{3gZ_i I_i}{\beta^3 \gamma^5 I_0} \frac{1}{Z_m^2} - \frac{\varepsilon_L^2}{\gamma^4} \frac{1}{Z_m^3} = 0, \qquad (1)$$

where $2Z_m$ is the length of the beam bunch at the longitudinal distance z travelled by the beam bunch center, g is a geometrical factor of order unity, Z_i and I_i are the initial beam pulse length and current, respectively, I_0 is the characteristic current given by $I_0=3.1\times10^7$(A/Z) amperes for ions of charge q=Ze and mass number A and $I_0=1.7\times10^4$ amperes for electrons, and ε_L is the longitudinal emittance of the beam.

---

* Research supported by the U.S. Department of Energy.

The transverse motion of the beam pulse can be analyzed by the transverse envelope equation under the assumption of a K-V distribution [4]

$$R'' + \kappa R - \frac{K}{R} - \frac{\varepsilon_T^2}{R^3} = 0 \qquad (2)$$

where R is the transverse envelope of the beam, κ is the periodic focusing function of the lens system, ε_T is the transverse emittance, and $K=(I/I_0)(2/\beta^3\gamma^3)$ is the generalized perveance.

The equations (1) and (2) are coupled by the g factor and the product KZ_m which is a constant during the compression due to the conservation of the total number of electrons. Solving these equations numerically yields the beam envelopes in both transverse and longitudinal directions. In our 5-m long channel, the solenoid lenses are spaced at intervals of S=13.6 cm. Fig. 1 shows these results for the initial beam parameters I_i=40 mA, T_i=50 ns, E_{head}=3.3 keV, E_{center}=5.0 keV. Under these conditions, the pulse length can be compressed by a factor of 3 when reaching the end of the channel. The current magnification factor is also around 3 in the compression.

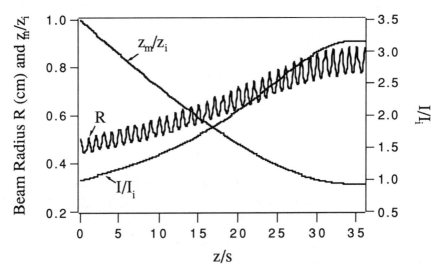

Fig. 1. Computer solution of the longitudinal compression.

2). Drift bunching of the beam with a uniform density distribution.

The beam with a parabolic density distribution is not common in the experiment. It is more desirable in practice to compress a beam with an initial uniform density distribution. In this case, initially, the space charge force is very strong at the two ends of the beam, and is very weak in its central region. The compression picture would be different from Fig. 1 and there is no analytical result available. We have run the "SLIDE" particle code[5] and the results are shown in Fig. 2. The density distribution is no longer a constant of motion. With an initial linear velocity tilt, the beam edges are hardly compressed due to the strong space charge force. These phenomena will be studied in detail in the experiment.

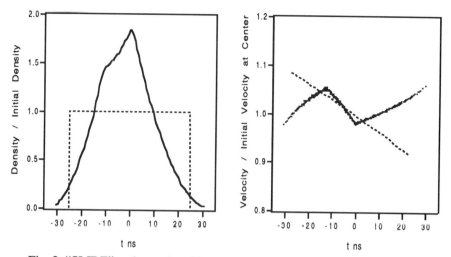

Fig. 2. "SLIDE" code results of beam compression with an initial (dotted lines) uniform density distribution and almost linear velocity tilt. The solid lines are the final density and velocity distributions after drift bunching in the 5-m long periodic solenoid focusing channel.

3). Experimental Design.

Fig. 3 shows a block diagram of the longitudinal compression experiment set up which consists of an electron beam injector, a periodic transport channel, and diagnostic chambers. The periodic transport channel is 5 m in length and 3.81 cm in diameter. It has 36 solenoid focusing lenses with the period length of 13.6 cm. The phase advance σ_0 is typically 72°. The tune depression σ/σ_0 could be as low as 0.1. The diagnostic chambers house all the equipments to measure the beam profile, energy, current, waveform, and emittance etc along the channel.

The electron beam injector can produce a beam for the compression experiment with the desired initial parameters: I_i=40 mA, T_i=50 ns, E_{head}=3.3 keV, E_{center}=5.0 keV, t_r < 1 ns. The configuration and performance characteristics of the injector is detailed in Section 4.

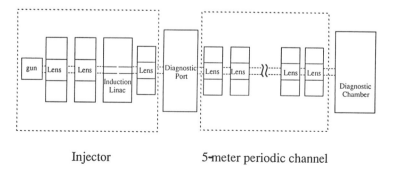

Fig. 3. Block diagram of the compression experiment set up.

LONGITUDINAL INSTABILITY

The longitudinal instability is an important and crucial issue for the development of induction linacs as drivers for heavy ion inertial fusion [6,7]. The special problems in this case are the space charge forces and the interaction between beams and induction gaps which are modeled by the complex impedances in this analysis.

1). Beam without energy spread.

For a mono-energetic beam transported in a channel with complex wall impedance $Z_w(s)$, the beam dynamics under density or velocity perturbation is governed by the dispersion equation [8]:

$$(s + ikv_0)^2 + k^2 \frac{gq\lambda_0}{4\pi\epsilon_0 m\gamma^5} + s\frac{q\lambda_0}{m\gamma^3} Z_w(s) = 0, \qquad (3)$$

where λ_0 and v_0 are the beam line charge density and velocity, and k and s are the wave number and complex frequency, respectively. The frequency ω_r and the growth rate ω_i of the perturbation can be found approximately as

$$\begin{cases} \omega_r \approx kc\beta \left[1 \pm \frac{1}{\gamma}\left(\frac{Kg}{2}\right)^{1/2} \left(1 - \frac{\beta\gamma^2}{g} X^*\right)\right] \\ \omega_i \approx \mp kc\beta^2 \gamma R^* \left(\frac{K}{2g}\right)^{1/2} \left(1 + \frac{\beta\gamma^2}{g} X^*\right) \end{cases}, \qquad (4)$$

under the condition that

$$\frac{2\beta\gamma^2}{g}(R^* + X^*) << 1,$$

where the wall impedance is approximated as $Z_w(s) \cong Z_w(-i\omega_0) = R(\omega_0) - iX(\omega_0)$ and normalized as

$$R^* = \frac{R(\omega_0)\Lambda}{Z_0}, \qquad X^* = \frac{X(\omega_0)\Lambda}{Z_0},$$

with $\omega_0 \cong kv_0$, $\Lambda = 2\pi/k$ and $Z_0 = 377\ \Omega$. Fig. 4 plots the growth rates of the slow waves for three different wall structures: resistive, inductive and capacitive walls. The growth rate can be significantly reduced by the capacitive wall, while the inductive wall increases the growth rate slightly.

When
$$\frac{2\beta\gamma^2}{g}(R^* + X^*) >> 1,$$

the approximate solution of Eq. (3) is

$$\begin{cases} \omega_r \approx kc\beta \left\{ 1 \pm \left[\frac{K\beta}{2} R^* \right]^{1/2} \right\} \\ \omega_i \approx \mp kc\beta \left[\frac{K\beta}{2} R^* \right]^{1/2} \end{cases} \quad (5)$$

This result is also shown in Fig. 4, where the curve B_1 is for a capacitive wall and B_2 is for a inductive wall..

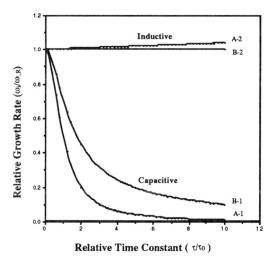

Fig. 4. Relative growth rate $\omega_i/\omega_{i,R}$ vs. the relative time constant τ/τ_0 for a wall modeled by R, L in series or R, C in parallel, where ω_i is the growth rate for inductive or capacitive wall, $\omega_{i,R}$ is the growth rate for a pure resistive wall, $\tau=RC$ for the capacitive wall, or $\tau=L/R$ for the inductive wall, and $\tau_0=1/\omega_0$. The curves A-1 and A-2 illustrate the results from Eq. (4) with the parameters $\beta=0.1$, R=100 Ω/m and $\omega_0=2\pi 10^8$ s^{-1}. The curves B-1 and B-2 represent the results of Eq. (5) where the parameters are $\beta=0.3$, R=300 Ω/m and $\omega_0=2\pi 10^6$ s^{-1}.

2). Beam with Energy Spread.

A beam with an initial energy spread is more stable due to Landau damping. The dispersion equation can be derived for this case as [9]:

$$1 = i \frac{q\lambda_0}{m\gamma^3} \left[\frac{kg}{4\pi \varepsilon_0 \gamma^2} + \frac{s}{k} Z_w(s) \right] \int \frac{df_0(v)}{dv} \frac{dv}{(s+ikv)} \quad (6)$$

where $f_0(v)$ is the distribution function at the meta-equilibrium state. The solution of this equation depends on the specific distribution function. It generally results in a stable region in the R*-X* plane. As an example, we suppose a Lorentz distribution:

$$f_0(v) = \frac{1}{\pi \alpha v_0} \frac{1}{\left(\frac{v - v_0}{\alpha v_0}\right)^2 + 1}, \quad (7)$$

where v_0 is the average velocity and α represents the degree of the velocity spread. The dispersion equation (6) for this distribution becomes

$$\frac{q\lambda_0}{m\gamma^3}\left[\frac{g}{4\pi\varepsilon_0\gamma^2} + \frac{s}{k^2}Z_w(s)\right] = \left[\frac{is}{k} + (i\alpha - 1)v_0\right]^2, \quad (8)$$

which leads to the boundary between the stable and unstable regions in the R*-X* plane, described by the relation

$$X^* = \left(\frac{R^*}{2\alpha} + \frac{1}{K\beta}\right)\left[\frac{Kg}{2\gamma^2} + \alpha^2 - \left(\frac{K\beta R^*}{K\beta R^* + 2\alpha}\right)^2\right]. \quad (9)$$

Fig. 5(a) shows the results for three different velocity spread parameters α at a fixed beam perveance K. The stable regions are bounded by the parabolic curves, and are mainly in the lower half plane, which is capacitive. Fig. 5(b) plots the stable regions for a fixed velocity spread but with different beam perveances K. It is evident that the space charge increases the instability dramatically.

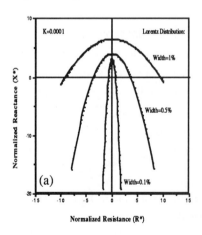
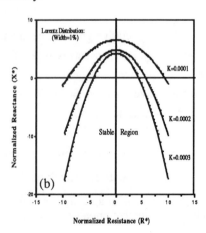

Fig. 5. (a). Stable regions in the R*-X* plane for three different velocity spreads of Lorentz distribution at a fixed beam perveance K. (b). Stable regions in the R*-X* plane for three different beam perveances at a fixed velocity spread.

3). Design for the resistive-wall instability experiment.

The experimental study of the longitudinal instability will start with an electron beam transported in a pure resistive-wall channel[10]. Fig. 6 is a block diagram of the experiment set up which consists of an electron beam injector, a resistive-wall transport channel, and the diagnostics.

The resistive-wall tube will be a one meter long glass tube coated with Tin Oxide in its inner surface. The resistance is designed to be 3 kilo-ohms per meter. The inner diameter of the tube is 3.81 cm. The tube is located inside and coaxial with a long solenoid which produces a uniform magnetic focusing field of around 100 Gauss. This field provides forces in transverse direction to balance space charge forces. An identical pipe with good conductivity will also be employed for comparison study.

The electron beam injector is the same as in the compression experiment. It provides a beam with the parameters of I=100 mA, E=2.5 keV, T=5 - 50 ns for the instability experiment.

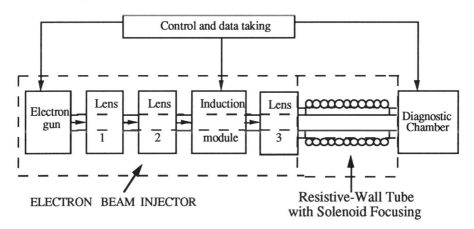

Fig. 6. Experimental set up for the resistive-wall experiment.

ELECTRON BEAM INJECTOR

An electron beam injector[11] has been built for the longitudinal compression and resistive-wall experiments. The injector consists of a variable-perveance gridded electron gun, three matching lenses and one induction acceleration module, as shown in Fig. 7.

The design of the variable-perveance gridded electron gun and its general performance characteristics were described in reference 12. Since then, many improvements have been made to the gun, including replacement of the ML-EE55 oxide cathode by a Y646B dispenser cathode assembly. The A-K gap has been modified to vary from 0.93 cm to 2.3 cm, resulting in a perveance of 0.22 to 1.35 $\mu AV^{-3/2}$. The gun can provide an electron beam with an adjustable energy up to 10 keV, and a current up to a few hundred milliamperes. The beam from the gun has a diameter of about 1 cm. The full beam emittance is measured by the pepper-pot method[13] to be 88 mm mrad with a standard deviation of 5 mm mrad. The beam pulse length is typically set to 50 ns and can be varied easily from a minimum of 2 ns up to hundreds of nanoseconds. The rise time of the square beam pulse is less than 1 ns. The beam pulse shape can be changed by changing the grid pulse shape. These features provide the flexibility of obtaining different beam parameters for the two experiments.

The matching lenses are made of ordinary solenoid with iron shielding. The axial magnetic field of the lense is a Gaussian shape with a peak value of about 100 Gauss. The equivalent lense width is about 4 cm. The lenses match the beam to the induction module and the experimental channel.

The design and performance characteristics of the induction acceleration module can be found in reference 14. The induction module consists of a single-turn primary around 2 ferrite cores with the beam completing the single-turn secondary. The gap voltage is time dependent from zero to 5 kV during each pulse. This is controlled by a PFN circuit, which results in an approximately t^2 shaped waveform with time. The induction linac employs a pseudospark switch to control the PFN operation. This component provides superior performance of the switch in the aspects of fast risetime, small jitter, current reversal capability, and long life time. In the compression experiment, the induction module will provide the necessary energy spread to the beam.

The injector system is undergoing the final test. The detailed results will be reported elsewhere.

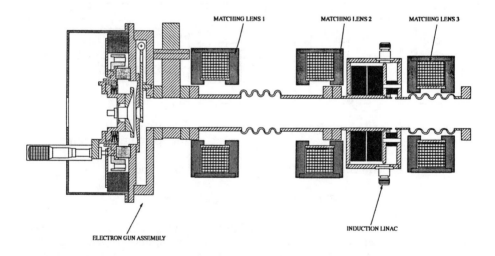

Fig. 7. Electron beam injector.

REFERENCES

1. T. Shea, E. Boggasch, Y. Chen, and M. Reiser, Proceedings of the IEEE Particle Accelerator Conference, p. 1049, Chicago Ill., March 20-23, 1989.
2. L. Smith, ERDA summer study for heavy ion inertial fusion (ed. R. O. Bangerter, W. B. Herrmannsfeldt, D. L. Judd, and L. Smith), LBL-5543 (1976), p.77.
3. D. Neuffer, IEEE Trans. on Nuclear Science, **26**(3), 3031, June 1979.

4. I. Kapchinsky and V. Vladimirsky, Proc. Conf. High Energy Accelerators, 2nd, p. 274, Geneva, 1959.
5. E. Henestroza, "SLIDE" code, LBL.
6. E. P. Lee and L. Smith, Proceedings of the 1990 Linear Accelerator Conference, Albuquerque, NM, September 10-14, 1990.
7. E. P. Lee, Proceedings of the 1981 Linear Accelerator Conference, Santa Fe, NM, October 19-23, 1981.
8. J. G. Wang and M. Reiser, Proceedings of the IEEE Particle Accelerator Conference, San Francisco, CA., May 6-9, 1991.
9. J. G. Wang and M. Reiser, CPB Tecknical Report #91-009, LPR, University of Maryland, May 1991.
10. J. G. Wang, M. Reiser, W. M. Guo, and D. X. Wang, to be published in Particle Accelerators.
11. J. G. Wang, D. X. Wang and M. Reiser, Proceedings of the IEEE Particle Accelerator Conference, San Francisco, CA., May 6-9, 1991.
12. J. G. Wang, E. Boggasch, P. Haldman, D. Kehne, M. Reiser, T. Shea, and D. X. Wang, IEEE Trans. Electron Devices, 37(12), pp. 2622-2628, Dec. 1990.
13. J. G. Wang, D. X. Wang, and M. Reiser, to be published in NIM, (A).
14. J. G. Wang, D. X. Wang, E. Boggasch, D. Kehne, M. Reiser, and T. Shea, NIM, (A) **301**, pp. 19-26, 1991.

STUDIES OF BRIGHT BEAM TRANSPORT BY THE LBL MFE GROUP*

O.A. Anderson, L. Soroka, and J.W. Kwan
Lawrence Berkeley Laboratory, Berkeley CA 94720, USA

ABSTRACT

We describe the coordination of theory and experiment in the study of bright beam systems by the LBL Magnetic Fusion Energy group. We have developed new analytic [1] and computational [2] tools for designing practical ESQ-focused systems, including the LBL CCVV (Constant Current Variable Voltage) prototype. This prototype has been tested both as a modular accelerator [3] and as part of a LEBT [4] for injection into an RFQ. We have also found these mathematical tools to be useful in setting up the bright-beam measurements on our test stand [5]. Our measurements showed no emittance growth in ESQ-focused 84 mA He^+ beams with emittances of 0.0075π mrad-cm (normalized rms). We discuss analytical work on emittance growth in segmented beams [6]; these results have been used as criteria for designing accelerators with multi-aperture sources [7]. The problem of emittance growth from beam size mismatch [8] is reviewed.

REFERENCES

[1] O.A. Anderson, Proc. 2nd European Particle Accelerator Conference, Nice, France, 1990; p. 1652.

[2] O.A. Anderson and L. Soroka., "ESQACL Code for Electrostatically Accelerated Beam with ESQ Focusing," Lawrence Berkeley Laboratory Report LBL-30488 (1991).

[3] O.A. Anderson, L. Soroka, et al., Nucl. Instr. and Meth. **B40/41**, 877 (1989).

[4] O.A. Anderson, L. Soroka, et al., Proc. 2nd European Particle Accelerator Conf., Nice, France, 1990; Editions Frontières, Gif-sur-Yvette, 1990, p. 1288.

[5] J.W. Kwan, et al., Proc. 14th Particle Accelerator Conference, San Francisco, 1991; preprint LBL-29974.

[6] O.A. Anderson, *Particle Accelerators* **21**, 197 (1987).

[7] O.A. Anderson, et al., Proc. 13th IAEA Plasma Physics and Controlled Fusion Conf., Washington DC (1990).

[8] O.A. Anderson, Proc. Int'l Symp. on Heavy Ion Fusion, M. Reiser, Ed., AIP Conf. Proc. **152**, 253 (1986).

O.A. Anderson and L. Soroka, "Emittance Growth in Intense Mismatched Beams," 1987 Particle Accelerator Conf., Washington DC; IEEE Cat. No. 87CH2387-9, p. 1043 (1987).

*Supported by U.S. DOE Contract DE-AC03-76SF00098.

HIGH-BRIGHTNESS H⁻ BEAM TRANSPORT USING ESQ LENSES*

S. K. Guharay, C. K. Allen, M. Reiser
Laboratory for Plasma Research
University of Maryland, College Park, MD. 20742

ABSTRACT

Development of an efficient low-energy beam transport section (LEBT) is an important issue in accelerator research. An experiment is undertaken to build up an electrostatic quadrupole (ESQ) lens system with an aim to transport a 30 mA, 35 kV H⁻ beam, with normalized beam brightness of $\sim 8 \times 10^{10}$ A/(m-rad)^2, over a length of about 30 cm and match it into a radio-frequency-quadrupole. Beam parameters of a Penning-Dudnikov type ion source at the AT-1 test stand at Los Alamos National Laboratory are used. The LEBT section consists of six electrostatic quadrupole lenses. The apparatus is designed by detailed computer code simulation of beam dynamics. An emittance growth of about 80% is predicted in transporting the beam through the ESQ system. The ESQ lens system is fabricated. Tests of the system are planned to validate the design parameters, voltage hold-off of the quadrupole assembly, etc. Critical issues on the design of the apparatus and its fabrication will be discussed.

INTRODUCTION

Charged particle beams with very high brightness are required in many modern applications. In today's as well as next generation's high energy colliders, e.g., Tevatron, SSC, NLC, etc., one of the vital requirements is to achieve luminosity of colliding charged particle beams of order 10^{31} cm^{-2}s^{-1} or higher; this demands that beam brightness also be very high.[1] H⁻ beams, with normalized brightness of $\gtrsim 10^{12}$ A/(m-rad)^2, are required in space defense for generation of intense neutral particle beams to probe any foreign objects. The important role of high intensity, high-brightness beams is also evident in the arena of heavy-ion fusion (HIF), free electron lasers, etc., and of late, in an attractive scheme for radio-active waste transmutation.[2]

The crux of the problem here is to obtain a high quality beam from an appropriate source with minimum distortions due to aberrations and nonlinear forces (fringe fields, image effects), and to preserve its quality, as much as possible, in the transport and acceleration chain. Alessi reviewed characteristics of beams, primarily H⁻ beams, from various types of sources.[3] Volume ionization as well as Penning-Dudnikov type sources are most suitable to obtain high brightness H⁻ beams. After extraction from the ion source the charged particle beam enters the low-energy beam transport (LEBT) section in an accelerator. The LEBT section

---

* Research supported by ONR/SDIO

acts as a phase-space transformer between the ion source and the acceleration chain in order to match the transported beam to the acceptance ellipse parameters of the first stage of acceleration, e.g. a radio-frequency-quadrupole (RFQ). Furthermore, it decouples the source from the acceleration chain for a clean operation of the accelerator. Designing an efficient LEBT system for high-brightness beams has been an issue of common interest in the accelerator community. This problem has been addressed in detail by Reiser,[4] who made a comparative study of various approaches for low-energy beam transport. It is important to identify experimentally various underlying issues governing the quality of the transported beam, especially, emittance growth, and compare the results with computer simulation predictions. A systematic effort in this regard is still missing. The present work emphasizes on this particular issue.

With the aim to transport a 30 mA, 35 kV H$^-$ beam over a distance of about 30 cm and focus it into an RFQ, an effort is made here to develop an electrostatic quadrupole (ESQ) lens system. One of the main objectives of this work is to develop computational tools to design a practical ESQ system and evaluate its confidence limit by comparing the computer predictions with measured values of beam parameters. The motivation for this work and fundamental design concepts of the ESQ system have been reported earlier.[5] Important features of the ESQ system are briefly reviewed in this article. The important addition here is an analysis dealing with the study of beam dynamics for three special types of input beam distributions – K-V, semi-Gaussian, and Gaussian. This yields an insight to the expected performance of the designed ESQ system.

THE LEBT SYSTEM

(a) Design of the apparatus and predictions of its performance by computer simulation:

As has been discussed in previous articles,[5] the design of the ESQ system is guided by detailed computer simulation studies. The input beam parameters conform to an H$^-$ source used in the BEAR experiment[6] at Los Alamos. A 30 mA, 35 kV round and diverging H$^-$ beam, with radius of 1 mm and normalized brightness of 8×10^{10} A/(m-rad)^2, is considered. First, a linear beam optics code is used to solve the K-V envelope equations; this defines the basic configuration of the ESQ lens system. With the aim to obtain a converging beam at the end of the LEBT channel of about 30 cm long, six lenses are cascaded along the line of beam propagation. At each stage, the beam envelope is constrained[5,7] so that distortions due to aberrations and image effects are minimized, and also, electrical breakdown is avoided. Geometrical parameters of the lens system are given in an earlier paper.[5] The beam envelope in Fig. 1 shows the K-V solution assuming an ideal hard-edge focusing function to drive the six ESQ lenses in the LEBT. In order to evaluate the field distribution in the ESQ system, equipotentials are mapped for the actual geometry of the lenses using a 3-D Laplace solver. This yields the spatial profile of the focusing function $\kappa(z)$, and fringe field.

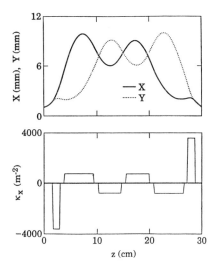

Figure 1: X and Y are beam excursions in x and y directions, respectively; z is the direction of beam propagation. $\kappa_x = -\kappa_y$.

With the above knowledge of the ESQ system, we proceed to learn about the beam dynamics, particularly in relation to the issue of emittance growth. A transfer matrix analysis (TMA) code, modified PARMILA,[8] is used here for this purpose. Here the applied field is modeled by a hard-edge function in the interior of the lens; fringe fields in the exterior region are obtained from the 3-D Laplace solver, as mentioned earlier, and included in the code following the technique of Matsuda and Wollnik.[9] The procedure has been described in detail earlier.[5] The main idea of the TMA code is to push particles through a chain of transfer matrices characterizing lens elements and drift spaces in the ESQ system. This code outputs spatial evolution of the beam geometry in both configuration and phase space as well as the beam emittance. The architecture of this code is well suited to determine the contribution of various nonlinear forces (e.g., chromatic aberration, fringe fields) in the predicted emittance growth. Analyses are done by turning on terms, one at a time, representing the respective nonlinear force fields in the transfer matrices.

Three types of distribution of the input H$^-$ beam are considered here for evaluating emittance growth in the designed ESQ system – a K-V type, a semi-Gaussian and a Gaussian.

Figure 2 shows results of the modified PARMILA code for a K-V type distribution of the input beam. It is noted that the beam at the output of the LEBT system ($z = 30.2$ cm) has acquired distortions. The emittance of the output beam is enhanced by a factor of 1.8; chromatic aberrations contribute about 90% of it.

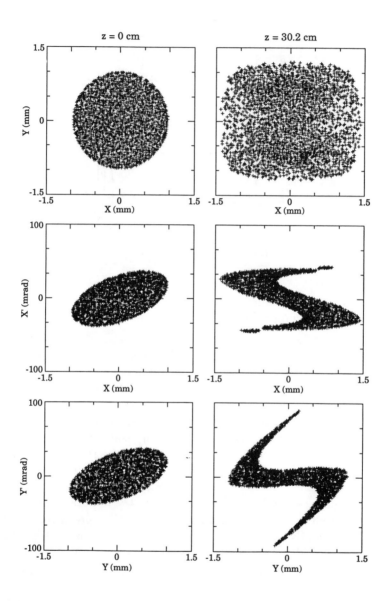

Figure 2: Modified PARMILA results with a K-V type input beam distribution. Left column shows the input beam geometry, and the right column shows the beam geometry at the output of the ESQ lens system.

Turning off the terms involving chromatic aberrations in the code, distortions in the output beam are reduced significantly – the elongated arms of the "S"-type output beam disappear. Going deeper into the analysis by evaluating emittance of the beam sequentially at each stage of the ESQ system, it is noted that the major enhancement of emittance dilution occurs at the second and fifth lenses of the ESQ assembly. Preliminary analyses of the computer simulation results suggest the following explanation for the underlying mechanism. It is evident in Fig. 1 that the first lens in the ESQ system is driven at a high voltage (~ 8 kV) to capture the highly divergent space-charge dominated H^- beam from the source (initial divergence angle of the beam envelope at the maximum radius is 20 mrad). Thus, a strong focusing force applied in one direction (say, along y) and a simultaneous defocusing force in the other orthogonal direction (along x) result in a large beam envelope in the defocusing plane at the entrance to the second lens. This gives rise to an enhancement in emittance in the defocusing direction at the second lens. Likewise, emittance is enhanced in the other orthogonal direction at the fifth lens, when the beam executes maximum excursion in that direction; the third and fourth lenses are effectively used as a FODO transport section to cover the length of the LEBT channel. This analysis points to the requirement of further study in terms of optimization of the ESQ system. A possible approach may be to split each of the second and fifth lenses into smaller ones and constrain the beam amplitude adiabatically by driving the lenses at higher voltage. In fact, the analysis could be done very economically using the K-V code, which delivers a good knowledge of the beam envelope.

The above analysis is repeated assuming a Gaussian distribution of the input beam. Figure 3 shows a comparison of the output beam geometry in configuration space for the two distributions of the input beam – K-V type (upper figure) and Gaussian with tail cut-off at 4σ (lower figure). In the Gaussian case, about 7% of total beam particles are seen on a circular annulus corresponding to the inner boundary of the ESQ lenses (second through fourth lens), and are thereby lost. The rapid blow-up of a Gaussian beam may be an artifact of excess energy in tail particles of the distribution function. The emittance growth in this situation is estimated to be a factor of about 3.3. Such an enhancement of emittance over the K-V case is due to particles located on the outer boundaries of phase space. These particles are essentially lost from the system or do not match with the desired acceptance ellipse parameters at the end of the LEBT. The phase space geometry of the core group of particles (contained within the dark boundary in the lower part of Fig.3) is compared with the K-V case by putting a scraper at $X = 15$ mm, $Y = 15$ mm in the Gaussian output beam. The result is shown in Fig. 4 – the solid boundary is for K-V case and the dashed boundary is for Gaussian. The phase space area occupied by the two boundaries appear to be similar. This suggests that the emittance of the output beam may be controlled by tailoring the distribution of the input beam, e.g., using a low-temperature beam. It is well understood that the two distributions, K-V and Gaussian, are too idealistic to

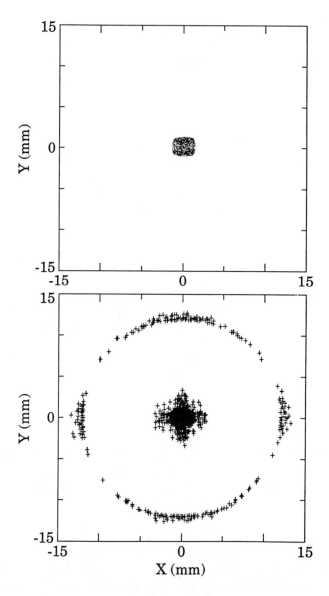

Figure 3: Output beam geometry in configuration space from modified PARMILA for the input beam distribution as: K-V (upper figure) and Gaussian (lower figure). Outer boundaries of the particles in the Gaussian case correspond to the location of the inner surface of ESQ lenses with larger aperture (radius = 12 mm); those particles are lost.

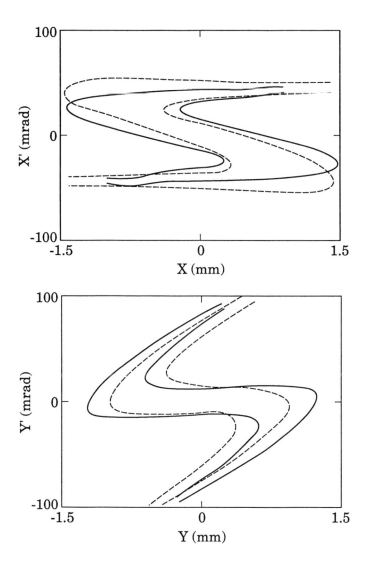

Figure 4: Boundary of phase-space geometry of the output beam from modified PARMILA: K-V case (solid line), core part of the Gaussian beam case (dashed line).

represent the input beam. Hence the aforementioned results predict performance of the ESQ system in two extreme situations. A semi-Gaussian distribution is a more natural candidate for the input beam distribution. Results with a semi-Gaussian distribution are found to be somewhat between the two extreme cases as discussed. Detailed analyses with various types of distributions are being carried out for an in-depth knowledge on this important problem and results will be reported elsewhere.

It is concluded from the aforementioned discussions that the beam excursion plays the key role in determining emittance. In a practical situation with a given lens system and setting of the lens power supplies, one control parameter for beam excursion may be the beam current. Figure 5 shows the K-V solution of beam envelope for beam currents at 25 mA, 30 mA, and 35 mA; the lens geometry and voltage on the ESQs are maintained the same. The beam excursion appears smaller at lower beam current. This gives an important guideline in setting the operating range of beam parameters for a given ESQ system.

(b) Experimental hardware:

The layout of essential components of the experiment is shown in Fig. 6. The main feature of the ESQ system is that it is adjustment-free. This demands a strong precision in machining individual components as well as assembling them together. The electrodes are fabricated in-house and their dimensional accuracy is measured to be within ±0.2 mil (average) of the designed value. Ceramic insulating balls, with precision in sphericity of ±0.025 mil, are used between electrodes of neighboring lenses as well as between an electrode and its adjacent ground plate. Special care has been taken to round off any sharp edges in the system.

The entire system will be pumped through a diagnostic box, carrying emittance scanners and Faraday cups. Adequate perforations are drilled on the ground plates as well as on spacers between the ground plates in order to avoid any localized pressure build-up and yet, maintain a good electrical shielding.

The LEBT system is currently being assembled and various tests are in progress. The experiment is planned to be done in two phases. First, the H$^-$ beam from the ion source needs to be characterized. Second, the LEBT will be interfaced with the ion source and diagnostic system, and beam transport experiments will be conducted.

CONCLUSION

Several key points of an ESQ beam transport design are addressed in the present work. Two codes, K-V and modified PARMILA, are mainly used to examine critically the factors influencing the beam emittance. In the present situation, chromatic aberration seems to be the primary contributor to the predicted emittance growth. The lens system is contoured following Laslett[10] and Dayton et al.[11] to reduce fringe field effects and thus minimize its contribution to the

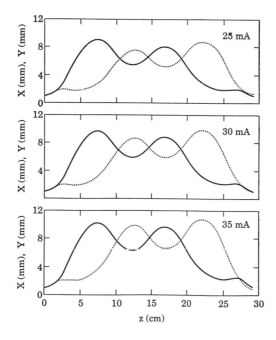

Figure 5: K-V envelope solutions for three different input beam current.

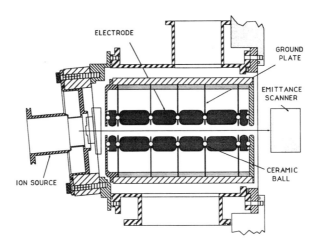

Figure 6: Layout of the experimental arrangement.

emittance growth; the computer simulation results support this principle in the present situation.

The design may be further optimized to reduce emittance growth by replacing the second and the fifth lenses in the ESQ system with a set of two smaller lenses. The envelope solution of K-V equations should then be constrained to remain near paraxial. Results with K-V, semi-Gaussian and Gaussian distributions of the input beam stress upon the issue of intrinsic quality of the beam in determining emittance growth in a beam transport system. Efforts are currently being made to address these issues in sufficient depth.

It may be remarked that none of the present studies includes any charge neutralization factor. Some preliminary analysis suggests that this may not be an important issue in the present case for a 35 kV H$^-$ beam, when a hydrogen gas pressure of about 10^{-4} Torr is considered. Detailed results will be reported elsewhere.

REFERENCES

1. J. T. Seeman, This Conference Proceedings; D. A. Edwards, ibid; N. M. Gelfand, ibid.
2. R. A. Jameson, ibid.
3. J. Alessi, ibid.
4. M. Reiser, Nucl. Instrum. & Meth. in Phys. Res. **B56/57**, 1050 (1991); Proc. 1988 Linear Accelerator Conf., 1988, p. 451; IEEE Trans. Nucl. Science **NS-32**, 2201 (1985).
5. S. K. Guharay, et al., Proc. Particle Accelerator Conf., San Francisco, CA, 1991 (to be published); Intense Microwave and Particle Beams II SPIE **1407**, 610 (1991).
6. P. G. O'Shea, et al., Nucl. Instrum. & Meth. in Phys. Res. **B40/41**, 946 (1989).
7. M. Reiser, et al., Microwave and Particle Beam Sources and Propagation SPIE **873**, 172 (1988).
8. C. R. Chang, et al., Intense Microwave and Particle Beams SPIE **1226**, 483 (1990).
9. H. Matsuda, and H. Wollnik, Nucl. Instrum. & Meth. **103**, 117 (1972).
10. L.J. Laslett, Design of Electrodes for the Single-Beam Ion Beam Experiment, LBL report HI-FAN-137.
11. I. E. Dayton, et al., Rev. Sci. Instrum. **25**, 485 (1954).

THE INFLUENCE OF THE BEAM PLASMA ON THE EMITTANCE OF INTENSE AND SPACE-CHARGE COMPENSATED BEAMS*

T. Weis

Institut für Angewandte Physik der Johann Wolfgang Goethe - Universität
Postfach 111 932, D-6000 Frankfurt am Main, Germany

ABSTRACT

Magnetic transfer lines are commonly used to allow for proper formation of ion beams between ion source and first accelerator structure as well as successive rf cavities. The absence of electric fields allows plasma build-up and the compensation of the space charge to a significant amount. The neutralization eases transverse focusing and allows for the stable transport of even very high beam currents, which under decompensated conditions would exceed the current limit of the channel. On the other hand the compensation process gives rise to internal self field nonlinearities of the beam resulting in transverse emittance degradation.

Fundamental aspects of space charge neutralization of positive ion beams will be discussed and causes for field nonlinearities followed by emittance degradation at low beam energies will be outlined. Numerical results of the beam plasma analysis will be given together with an overview on the pros and cons of the use of space charge compensated transport.

INTRODUCTION

Modern accelerator sytems often require high beam current at minimum emittances to allow for high beam luminosity at the high energy end. Due to high space charge forces the main fraction of beam quality degradation occurs in the low energy part of the system and is due to field nonlinearities produced by the accelerator and transport structures but also depends on the internal properties of the beam space charge forces. Careful design of all elements is therefore strongly recommended to keep the emittance growth low.

The Low Energy Beam Transport (LEBT)-line between ion source and first accelerator part is a crucial part of the whole injector system since the current limits of Radio-Frequency-Quadrupole-(RFQ)-resonators can hardly be obtained by transport systems without the use of space charge compensation[1]. Therefore magnetic focusing systems (solenoids and quadrupoles) are mainly used to allow for the trapping of electrons in the transverse space charge potential of the positive ion beam. Thus the space charge forces of the beam can be lowered significantly.

The dynamic equilibrium of such a beam plasma under compensating conditions is generated by manifold processes and the main contributors are summarized in table I.

Low energy ion beams with currents in the range between mA and hundreds of mA have particle densities, which are even at high pressures (10^{-3} hPa) at least an order of magnitude smaller than the density of the surrounding residual gas. Collisions between beam ions and gas atoms give rise to ionization of the residual gas particles. Depending on the location and energy of the created

---

*Work supported by BMFT under contract no. 06 OF 186I

electron, the negative charged particle can be trapped by the positive transverse space charge potential of the beam or immediately hit the chamber walls. The created slow residual gas ions (P1 and P2 in table I) are repelled from the beam and are lost transversely in a time scale of less than 1 μsec.

Table I Summary of the main particle interactions in a beam plasma
(positive ion beam in residual gas)

P1	$I^+ + X^0 \longrightarrow I^+ + X^+ + e^-$	ionization of the residual gas atom X
P2	$I^+ + X^0 \longrightarrow I^0 + X^+$	charge transfer
P3	$I^+ + X^0 \longrightarrow I^{2+} + X^+ + 2e^-$	P1 + successive ionization of I^+
P4	$I^0 + X^0 \longrightarrow I^0 + X^+ + e^-$	ionization of X by fast neutral atoms
P5	$I^+ + e^-_{slow} \longrightarrow I^+ + e^-_{fast}$	heating of compensating electrons
P6	$I^+ + X^0 \longrightarrow I^+_{scat} + X^0_{scat.}$	scattering of beam ions
P7	$X^+ + e^- \longrightarrow X^0 + \gamma$	electronic recombination; light emission
P8	$X^+ + 2e^- \longrightarrow X^0 + e^-$	3-body electronic recombination
P9	$X^0 + e^- \longrightarrow X^+ + 2e^-$	ionization of X by electrons
P10	$e^- + e^- \longrightarrow e^-_{scat.} + e^-_{scat.}$	scattering of electrons

The degree of compensation of the space charge potential is mainly influenced by the properties of the electron cloud. Process P10 in table I gives rise to a very fast thermalization of the velocity distribution of the electrons. The electron temperature depends also on the possible heating of the distribution via Coulomb collisions with the beam ions or plasma instabilities [2,3,4]. The cross sections σ of the processes P1 and P2 are in the order of 10^{-17}-10^{-15} cm^2. A more comprehensive overview on the process of plasma build-up is given by Holmes [2].

The finite electron temperature is one of the reasons why a fully compensated beam (no remaining electric space charge field inside the beam volume) cannot be reached and in addition, it causes nonlinearities of the residual space charge fields under partly compensating conditions (see below).

EMITTANCE GROWTH IN LEBT SECTIONS

Sources of transverse rms emittance growth in uncompensated beams have been clearly identified [5,6,7], namely aberrations caused by nonlinear fields of the transport channel and charge density redistribution of the beam, thus minimizing the internal field energy of the ions in their own space charge field until a homogeneous density has been reached. The original idea[5] has been extended by

Reiser for off-centered and mismatched beams[8] and has been proved experimentally[9]. The redistribution of the charge density causes an energy transfer from field energy to transverse kinetic energy and an increasing rms emittance. The process is adiabatic and occurs once, if only linear external fields exist.

At first sight this looks quite favorable in the neutralized beam case. First, the beam perveance and therefore the nonlinear field energy is much smaller and second, smaller beam radii can be achieved and therefore the influence of external field nonlinearities on the emittance is low[10]. On the other hand, charge density redistributions in partially space charge compensated beams means rearrangement of both ions and electrons to give a linear self consistent space charge field. Unfortunately the compensating electrons in the case of a positive ion beam are not cold enough to be exactly pinned to the ions. The thermal velocity distribution of the electrons can easily cause broader electron distributions in space compared to a homogeneous ion density distribution, resulting in a net negative space charge outside the beam. This has been demonstrated in Frankfurt using a transverse electron beam probe ($1\,\mu A$, $1\,keV$) for the determination of the transverse net charge density distribution[11].

THE ELECTRIC SELF FIELD OF THE BEAM PLASMA

For the evaluation of possible quality loss of a partially compensated ion beam the transverse electric self field of the beam plasma is of major importance. All charged particles present in the plasma contribute to the self field and the corresponding space charge potential can be obtained from Poissons equation.

The densities however depend on the potential in a complicated way and on the energy balance of the system[2]. Nevertheless the following statements can be made. The time scale of the thermalization of the electron velocity distribution is short compared to the time scale of all other processes in the plasma. We therefore assume a Maxwellian velocity distribution with a temperature T and the local electron density n_e as a function of the radial distance r from the axis is given by (assuming radial symmetry and a dc beam)

$$n_e(r) = n_{e0}\, e^{-e\varphi(r)/kT} * \frac{\varphi_R - \varphi(r)}{\varphi_R} \qquad (1)$$

with n_{e0} denoting the electron density for $r = 0$. φ_R is the overall space charge potential between beam axis and beam pipe and φ the potential as a function of r. The fraction takes into account that electron energies exceeding $e\varphi_R$ are not possible as considered by Holmes[2].

For residual gas pressures of the order 10^{-4} hPa and less, the influence of the created residual gas ions on the potential can be neglected. In the high pressure region and especially for almost fully compensated beams, the residual gas ion density can easily exceed the density of the beam ions. With σ_r as the cross section for the production of residual gas ions via the processes P1 and P2 of table I, v_i the beam velocity, n_0 the residual gas density, m_r the mass of the residual gas ions and $n_i(r)$ the ion beam density inside the beam volume, the density of the residual gas ions $n_r(r)$ is given by [2]

$$n_r(r) = \frac{n_0 v_i \sigma_r \sqrt{m_r}}{r \sqrt{2e}} * \int_0^r \frac{n_i(r') \, r' \, dr'}{\sqrt{\varphi(r) - \varphi(r')}} \,. \qquad (2)$$

Here free ion fall of the residual gas ions in the transverse beam potential has been assumed. For a given density of the ion beam $n_i(r)$ and a long cylindrical beam in a circular pipe, Poissons equation and the equations 1-2 can be combined to give

$$\frac{d^2\varphi}{dr^2} + \frac{1}{r}\frac{d\varphi}{dr} = -\frac{e}{\varepsilon_0} * \left(n_i(r) + n_r(r) - n_e(r) \right) \,. \qquad (3)$$

Equation 3 is of implicit form, because n_r and n_e are all functions of the potential φ itself. Additionally the set of equations 1-3 is not complete. Due to the missing equation for the energy balance, the electron temperature T as well as the density of the electrons at the beam axis are a priori not known and numerical solutions of equation 3 give solutions for $\varphi(r)$ depending on the choice of T and n_{e0}. From the solutions one has to choose those which are physically reasonable.

For high beam currents, the resulting densities of beam ions and electrons are high enough to result in a Debye length significantly smaller than the transverse beam dimensions. The beam plasma therefore will be neutral and the violation of the plasma neutrality will only occur at the beam edge in a positively charged plasma sheat with a thickness of the order of the Debye length. If one asssumes plasma neutrality at the beam axis, n_{e0} can be obtained numerically by iterative methods.

For a 10 keV, 1.5 mA He$^+$ ion beam equation 3 has been solved numerically assuming plasma neutrality at r = 0 (see figs. 1-7). The electron temperature T has been chosen such that the derivative of the potential φ with respect to r is zero at the beam pipe wall. A higher temperature T will result in an increased amount of electrons occupying the volume between beam and pipe and in a negative slope of $\varphi(r)$ near the wall causing more electrons to get lost until the equilibrium has been reached. A smaller value of T results in a lower degree of compensation. The comparison of the numerical results with measured values of the degree of compensation[12,13] however indicates that this solution seems to be reasonable only for low degrees of compensation for residual gas pressure lower than 10^{-5} hPa.

Two types of ion beams have been analyzed: a beam with homogeneous space charge density and a radius of 15 mm in a cylindrical pipe with radius R = 50 mm, and a beam with Gaussian transverse density profile (second moment of particle density distribution $<r^2>$ = 10 mm, R = 50 mm); $\sigma_r = 5.8 * 10^{-16}$ cm^2.

Figure 1 shows the numerical solution for the normalized electric self field E(r) and potential $\varphi(r)$ for the uncompensated beam with Gaussian profile. The self field is highly nonlinear and will cause rearrangement of the density distribution and rms emittance growth of the beam. Note that the self field E(r) will be linear for the beam with constant density under uncompensated conditions. This beam can be transported in linear external focusing fields without degradation of beam quality.

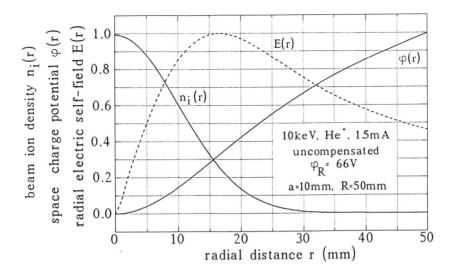

Fig. 1. Numerical solution of equation 3 for the uncompensated beam with Gaussian density profile. The normalized ion beam density n_i, electric self field E and space charge potential φ ($\varphi(r)/\varphi_R$) versus r are given. φ_R = 66 V.

In the compensated case the properties of the electric self field E(r) are totally different. For a Gaussian profile (fig. 2) and a residual gas pressure of 10^{-5} hPa the electric self field is linear to a great extent and shows nonlinearities only for the outer parts of the beam. The transverse distributions of the residual gas ions and the compensating electrons are also of Gaussian shape (fig. 3). The degree of compensation $f_{total} = 1 - \varphi_{R, comp}/\varphi_{R, uncomp}$ was found to be 0.92, which was in excellent agreement with the experiment[12]. The density of the residual gas ions is rather small. In the high pressure regime (10^{-3} hPa), the self field E(r) is also mostly linear (fig. 4) and the particle densities have Gaussian profiles (fig. 5), a result obtained earlier by Holmes[2]. The residual gas density in the center of the beam is about 50% of the ion beam density, thus contributing to the effective space charge density in a not negligable way. In order to obtain plasma neutrality at r = 0, the electron density is the sum of ion beam and residual gas density.

For a constant density profile of the ion beam, however, the numerical results show a highly nonlinear behaviour of the internal self field E under compensated conditions at 10^{-5} hPa (fig. 6). Due to the finite temperature the electron density is decreasing with radius and gives rise to a negative net charge density n_n outside the beam volume (fig. 7).

EMITTANCE GROWTH OF COMPENSATED BEAMS

Rearrangement of the space charge distribution for a compensated beam therefore does not necessarily end up in a homogeneous ion distribution. Moreover the numerical results indicate that Gaussian shaped density profiles of the beam ions seem to be self consistent distributions under compensating conditions. A compensated beam therefore tends to rearrange such that the potential energy of the ions (field energy) in the beam plasma reaches its

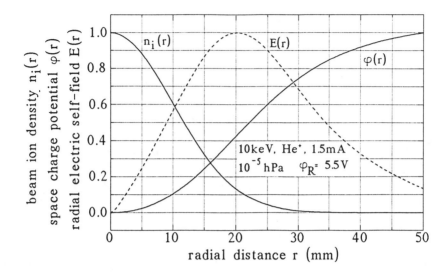

Fig. 2. Numerical solution of equation 3 for the compensated beam with Gaussian density profile. The normalized ion beam density n_i, electric self field E and space charge potential φ versus r are given. $\varphi_R = 5.5$ V for a residual gas pressure of 10^{-5} hPa. e^--temperature $kT = 1.53$ eV

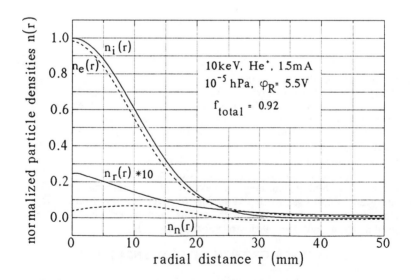

Fig. 3. Numerical solution of equation 3 for the compensated beam with Gaussian density profile. The normalized ion beam density n_i, electron density n_e, residual gas ion density n_r (enlarged by a factor of ten) and net charge density n_n as a function of r are given. $\varphi_R = 5.5$ V, $kT = 1.53$ eV.

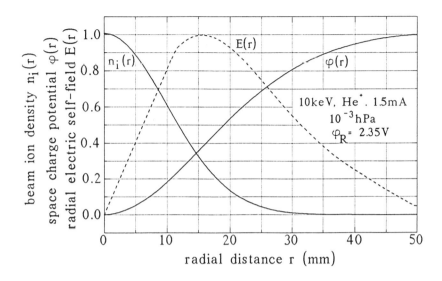

Fig. 4. Numerical solution of equation 3 for the compensated beam with Gaussian density profile. The normalized ion beam density n_i, electric self field E and space charge potential φ versus r are given. $\varphi_R = 2.35$ V for a residual gas pressure of 10^{-3} hPa. e^--temperature $kT = 0.86$ eV.

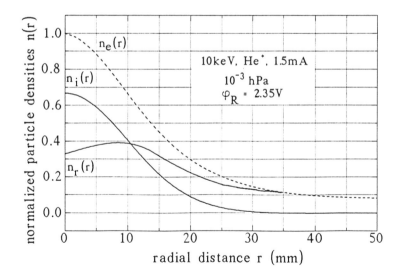

Fig. 5. Numerical solution of equation 3 for the compensated beam with Gaussian density profile. The normalized ion beam density n_i, electron density n_e and residual gas ion density n_r as a function of r are given. $\varphi_R = 2.35$ V. $kT = 0.86$ eV.

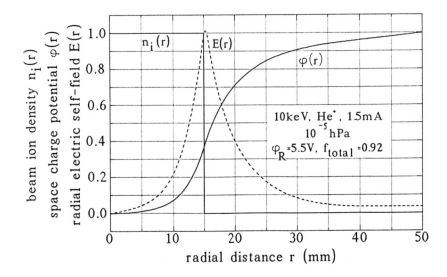

Fig. 6. Numerical solution of equation 3 for the compensated beam with homogenious density profile. The normalized ion beam density n_i, electric self field E and space charge potential φ versus r are given. $\varphi_R = 5.5$ V for a residual gas pressure of 10^{-5} hPa. e^--temperature kT = 4.55 eV.

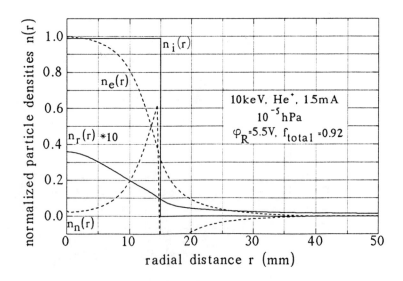

Fig. 7. Numerical solution of equation 3 for the compensated beam with homogenious density profile. The normalized ion beam density n_i, electron density n_e, residual gas ion density n_r (enlarged by a factor of ten) and net charge density n_n as a function of r are given. $\varphi_R = 5.5$ V, kT = 4.55 eV.

minimum value (Gaussian shape of electrons, ions and residual gas ions) causing emittance growth in the same way as an uncompensated beam does. This has been shown experimentally in Frankfurt[13].

In contrast to the uncompensated case this process is not adiabatic. The loss and the creation of electrons is a dynamic process and highly influenced by energy exchange between external fields and the beam plasma. The variation of the space charge neutralization rate, caused by changing residual gas pressure or even by varying beam radii[13] or the local decompensation due to electric fields is a continuous source for emittance degradation and measurements at GSI with a partially compensated 190 keV Ar^+ beam transported in a magnetic quadrupole channel have indeed shown significant increase of emittance compared to the unneutralized transport[14].

THE PROS AND CONS OF COMPENSATED BEAM TRANSPORT

Our experimental and theoretical results[12,13] and the numerical results presented in this paper have shown that emittance degradation of partially space charge compensated beams is originated from unlinearities of the residual self field of the beam. The rates of emittance increase due to the rearrangement of ions, electrons and residual gas ions are rather small (high degree of compensation and low field energy) but these rearrangements may occur more than once in the system depending on variations of gas pressure, beam radii and structure length resulting in an substantial amount of quality loss.

The results indicate that after an initial density redistribution the uncompensated beam transport in the absence of aberrative elements is the only way to avoid further emittance degradation.

Since the residual gas pressure behind the ion source extraction region is often too high to avoid space charge neutralizing effects, the use of electric transport elements seems to be a way out. Voltage hold-off and the necessity to provide high focusing strengths for unneutralized high perveance beams are the major problems.

Since the beam is unneutralized, the time structure of the beam plays a less significant role. Beam behaviour and particle dynamics can be described very accurately by numerical simulations. Apart from the emittance increase right after the extraction system, the growth of emittance is governed only by the aberrations of the system and can be kept small by sophisticated design methods.

The initial emittance increase after extraction is much smaller for a partly compensated beam. Depending on the degree of compensation and the variation along the beam path however, the quality of the beam will be degraded and the beam will redistribute in front of an rf accelerator e.g. causing additional emittance growth.

Therefore each individual case whether to choose an electric or magnetic transport system has to be decided on its own merits in order to provide a beam transport with minimum emittance growth.

ACKNOWLEDGEMENTS

The author wishes to express his great appreciation to Martin Reiser and his group for the organization of the encouraging workshop.

REFERENCES

1. T. Weis, Proc. 1990 Europ. Part. Acc. Conf., (Editions Frontieres, Gif-sur-Yvette, 1990) 181
2. A.J.T. Holmes, Phys. Rev. A19 (1979) 389
3. T. Weis, R. Dölling, P. Groß, H. Klein, J. Wiegand, Nucl. Instr. Meth. A278 (1989) 224
4. M.V. Nezlin, Plasma Phys. 10 (1968) 337
5. J. Struckmeier, J. Klabunde and M. Reiser, Part. Accel. 15 (1984) 47
6. T. Wangler, K.R. Crandall, R.S. Mills and M. Reiser, IEEE Trans. Nucl. Sci. NS-32 (1985) 2196
7. I. Hofmann and J. Struckmeier, Part. Accel. 21 (1987) 69
8. M. Reiser, J. Appl. Phys. 70 (1991) 1919
9. D. Kehne, M. Reiser and H. Rudd, Proc. 1991 Part. Accel. Conf., San Francisco, in print
10. T. Wangler, 1988 Lin. Acc. Conf., Cebaf-Report 89-001 (1989) 211
11. P. Groß, R. Dölling, T. Weis, J. Pozimski, J. Wiegand and H. Klein, Proc. 1990 Europ. Part. Accel. Conf., (Editions Frontieres, Gif-sur-Yvette, 1990) 806
12. T. Weis, J. Wiegand, R. Dölling, J. Pozimski, H. Klein and I. Hofmann, Proc. 1990 Lin. Accel. Conf., LANL-Report LA-12004-C (1991) 358
13. T. Weis, J. Wiegand, R. Dölling, J. Pozimski, F. Fenger and H. Klein, Proc. 1990 Europ. Part. Accel. Conf., (Editions Frontieres, Gif-sur-Yvette, 1990) 809
14. J. Klabunde, A. Schönlein, 1986 Lin. Acc. Conf., SLAC-Report 303 (1986) 296

ULTRA-INTENSE LASER INTERACTIONS WITH BEAMS AND PLASMAS

Phillip Sprangle and Eric Esarey
Beam Physics Branch, Plasma Physics Division
Naval Research Laboratory, Washington, DC 20375-5000

ABSTRACT

The nonlinear interaction of ultra-intense laser pulses with electron beams and plasmas is rich in a wide variety of new phenomena. Advances in laser science have made possible compact terawatt lasers capable of generating subpicosecond pulses at ultra-high powers (≥ 1 TW) and intensities ($\geq 10^{18}$ W/cm^2). These ultra-high intensities may have applications in the areas of advanced accelerators and high-brightness electron beams. This paper briefly addresses a number of relevant phenomena, including (i) laser excitation of large amplitude plasma waves (wakefields), (ii) relativistic optical guiding of laser pulses in plasmas, (iii) optical guiding by preformed plasma channels, and (iv) cooling of electron beams by intense lasers.

I. INTRODUCTION

Advances in laser technology have made possible compact terawatt laser systems with high intensities ($\geq 10^{18}$ W/cm^2), modest energies ($\lesssim 100$ J) and short pulses ($\lesssim 1$ psec).[1,2] This new class of lasers is referred to as T^3 (Table-Top-Terawatt) lasers. The availability of T^3 lasers has made possible experiments in a new ultra-high intensity regime. Previous laser interaction studies have been limited, for the most part, to relatively modest intensities. At ultra-high intensities, the laser-electron interaction becomes highly nonlinear and relativistic, thus resulting in a wide variety of new and interesting phenomena.[3-13] These phenomena include: (i) laser excitation of large amplitude plasma waves (wakefields),[3-5] (ii) relativistic optical guiding of laser pulses by plasmas,[5-7] (iii) optical guiding by preformed plasma channels,[8] and (iv) the cooling of electron beams by intense lasers.[13] These phenomena may have important applications ranging from advanced ultra-high gradient accelerators to advanced sources of ultra-short wavelength coherent radiation. This paper briefly discusses some of the salient features of these phenomena.

An important parameter in the discussion of ultra-intense laser interactions is the unitless laser strength parameter, a_o, where $a_o = |e|A_o/m_o c^2$ is the normalized peak amplitude of the laser vector potential, A_o. The laser strength parameter is related to the power, P_o, of a linearly polarized laser by

$$P_o[\text{GW}] = 21.5(a_o r_o/\lambda_o)^2, \qquad (1)$$

where r_o is the spot size of the Gaussian profile, λ_o is the laser wavelength, and the power is in units of GW. Physically, $a_o \geq 1$ implies that the electron quiver motion in the laser field is highly relativistic and nonlinear. This may be seen from conservation of canonical transverse momentum. In the 1-D limit $a_o = \gamma\beta_\perp$, where γ is the relativistic factor and $\beta_\perp = v_\perp/c$ is the electron quiver velocity; hence, $a_o \gg 1$ implies $\gamma \gg 1$. The peak laser electric field amplitude, E_o, is related to the laser parameter, a_o, by

$$E_o[\text{MeV/m}] \simeq 3 \times 10^6 \, a_o/\lambda_o[\mu m]. \qquad (2)$$

For $\lambda_o = 1$ μm and $a_o \geq 1$, the laser electric field exceeds the Coulomb field associated with the hydrogen atom. In terms of the laser intensity ($I_o = 2P_o/\pi r_o^2$), the quantity a_o is given by

$$a_o = 0.85 \times 10^{-9} \lambda_o[\mu m] I_o^{1/2}[\text{W/cm}^2], \qquad (3)$$

where λ_o is in units of μm and I_o is in units of W/cm^2. Highly relativistic electron motion ($a_o \geq 1$) requires laser intensities greater than 10^{18} W/cm^2 for wavelengths of ~ 1 μm. Such intensities are now available from compact, T^3 laser systems.

A. Compact Terawatt Lasers

The T^3 laser system is based on the technique of chirped-pulse amplification (CPA), first applied to solid-state lasers in 1985.[1] The CPA technique allows for ultra-short (≤ 1 ps) pulses to be efficiently amplified in solid-state media, such as Nd:Glass, Ti:Sapphire and Alexandrite.[1,2] In the T^3 laser, a low energy pulse from an ultra-short pulse, mode-locked oscillator is passed through an optical fiber to produce a linear frequency chirp. The linear frequency chirp allows the pulse to be temporally stretched by a pair of gratings. The stretched pulse is amplified to moderate energies in the solid-state regenerative and single pass amplifiers. The amplified pulse is now compressed by a second matched pair of gratings. Compression of the chirped, long duration pulse is accomplished after it has been amplified, thus avoiding undesirable high field effects, e.g., self-focusing, in the solid-state medium. The CPA method is schematically shown in Fig. 1. This method has been applied to compact systems (T^3 lasers) to produce picosecond pulses in the 1-10 TW range.[1,2] Efforts are also underway to apply the CPA method to large-scale systems with the goal of producing laser pulses of extremely high energies (> 100 J) and powers (> 100 TW).[14] In an alternative technology, large-scale KrF excimer lasers systems can directly amplify a single short pulse to terawatt power levels.[15]

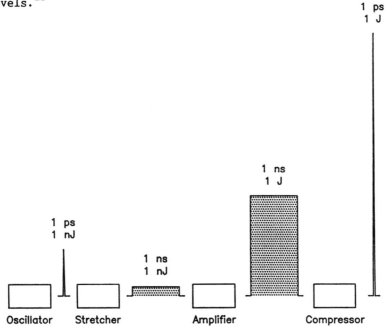

Fig. 1 Schematic of the chirped-pulse amplification (CPA) method used in T^3 laser systems.

II. INTENSE LASER INTERACTION PHENOMENA

A. Laser Wakefield Excitation

As an intense laser pulse propagates through an underdense plasma, $\lambda_o^2/\lambda_p^2 \ll 1$, where $\lambda_p = 2\pi c/\omega_p$, $\omega_p = (4\pi|e|^2 n_o/m_o)^{1/2}$ is the plasma frequency and n_o is the ambient electron density, the ponderomotive force associated with the laser pulse envelope, $F_p \sim \nabla a^2$, expels electrons from the region of the laser pulse. If the pulse length is approximately equal to the plasma wavelength, $c\tau_L \simeq \lambda_p$, the ponderomotive force excites large amplitude plasma waves (wakefields) with phase velocities equal to the laser pulse group velocity,[3-5] [see Fig. 2]. The maximum wakefield amplitude generated by a linearly polarized laser pulse of amplitude a_o, in the 1-D limit $r_o^2 \gg \lambda_p^2$, is[5]

$$E_{max}[\text{GeV/m}] = 3.8 \times 10^{-8} (n_o[\text{cm}^{-3}])^{1/2} \frac{a_o^2}{(1 + a_o^2/2)^{1/2}}, \quad (4)$$

where the maximum gradient, E_{max}, is in GeV/m and the plasma density, n_o, is in cm^{-3}. Typically, E_{max} is a few orders of magnitude greater than the accelerating gradients in conventional linear accelerators.

The accelerating gradient associated with the wakefield can accelerate a trailing electron beam, i.e., such as in the laser wakefield accelerator (LWFA), as shown in Fig. 2. In the absence of optical guiding, the interaction distance, L_{int}, will be limited by diffraction. The interaction distance is $L_{int} \simeq \pi Z_R$, where $Z_R = \pi r_o^2/\lambda_o$ is the vacuum Rayleigh length. The maximum energy gain of the electron beam in a single stage is $\Delta W = E_{max} L_{int}$ which, in the limit $a_o^2 \ll 1$, may be written as

$$\Delta W[\text{MeV}] = 580 \, (\lambda_o/\lambda_p) P_o[\text{TW}]. \quad (5)$$

As an example, consider a τ_L = 1 psec linearly polarized laser pulse with P_o = 10 TW, λ_o = 1 μm and r_o = 30 μm (a_o = 0.72). The requirement that $c\tau_L = \lambda_p$ implies a plasma density of $n_o \sim 1.2 \times 10^{16}$ cm^{-3}. The wakefield amplitude (accelerating gradient) is E_{max} = 2.0 GeV/m and the interaction length is L_{int} = 0.9 cm. Hence, a properly phased tailing electron bunch would gain an energy of ΔW = 18 MeV. The interaction length, and consequently the electron energy gain, may be greatly increased by optically guiding[5-8] the laser pulse in the plasma.

LASER WAKEFIELD ACCELERATOR

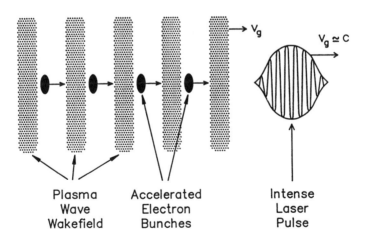

Fig. 2 Schematic of the laser wakefield accelerator (LWFA). The radiation (ponderomotive) force of an intense laser pulse drives plasma waves (wakefields) with phase velocities ~ c which trap and accelerate trailing electron bunches.

The self-consistent evolution of the nonlinear plasma wave has been studied numerically in the 1-D limit.[5] Figure 3 shows the plasma density variation $\delta n/n_o = n/n_o - 1$ and the corresponding axial electric field E_z for a laser pulse envelope with $c\tau_L = \lambda_p$. In this figure $c\tau_L = \lambda_p = 0.03$ cm, $\lambda = 10$ µm, and $a_o = 2.0$. The steepening of the electric field and the increase in the period of the wakefield are apparent for the highly nonlinear situation shown in Fig. 3.

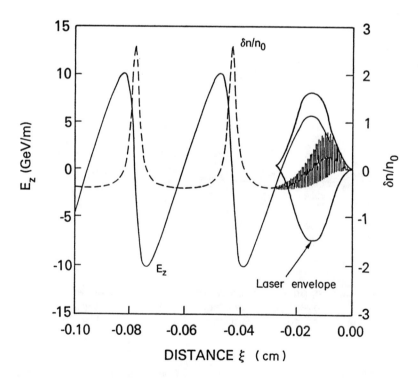

Fig. 3 Density variation $\delta n/n_o = n/n_o - 1$ and axial electric field E_z in GeV/m for a laser pulse with $a_o = 2$ and $c\tau_L = \lambda_p = 0.03$ cm.

B. Relativistic Optical Guiding

For a laser pulse to propagate in a plasma beyond the limits of vacuum diffraction, i.e., distances large compared to Z_R, some form of optical guiding is necessary. For sufficiently powerful, long laser pulses, diffraction can be overcome by relativistic effects and the laser pulse can be optically guided in the plasma.[5-7] A nonlinear analysis of intense laser pulse propagation in underdense plasmas indicates that the index of refraction, η_R, characterizing the laser pulse evolution is given by[5]

$$\eta_R = 1 - (1/2)(\omega_p/\omega)^2 = 1 - (1/2)(\lambda_o/\lambda_p)^2 \gamma^{-1} n/n_o, \qquad (6)$$

where n is the perturbed electron density. For a long laser pulse (long rise time), $c\tau_L \gg \lambda_p$, it may be shown that $\gamma^{-1} n/n_o \simeq (1 + a_o^2/2)^{-1/2}$. A necessary requirement for optical guiding is that the refractive index have a maximum on axis, $\partial \eta_R/\partial r < 0$. This is the case for a long laser pulse with an intensity profile peaked on axis, $\partial a_o^2/\partial r < 0$. Analysis of the wave equation with the index of refraction given by Eq. (6) indicates that the main body of a long laser pulse will be optically guided, provided the laser P_o exceeds a critical threshold, $P_o > P_{crit}$, where[6,7]

$$P_{crit}[GW] = 17.4(\lambda_p/\lambda_o)^2. \qquad (7)$$

As an example, for a plasma of density $n_o = 10^{19} cm^{-3}$, $\lambda_p = 11$ µm, and laser wavelength of $\lambda_o = 1$ µm, the critical laser power is $P_{crit} = 2.1$ TW. For sufficiently short pulses, $\tau_L < 1/\omega_p$, the plasma has insufficient time to respond to the laser pulse. The modification of the refractive index occurs on the plasma frequency time scale, not on the laser frequency time scale. Therefore, if the laser pulse duration is comparable to or shorter than a plasma period, i.e., $\tau_L \lesssim 1/\omega_p$, relativistic optical guiding does not occur.[5]

C. Optical Guiding by Density Channels

To optically guide short intense laser pulses in plasmas preformed plasma density channels can be used.[8] Conventional optical light pipes consist of fibers having an index of refraction which varies as the square of the radial position. Optical light guides can be formed within a plasma by creating a hollow density channel. The plasma channel can be formed by propagating either a low current electron beam or a low intensity laser pulse through the plasma. In either case the plasma is expelled by the beam, leaving a density depression which can guide the laser pulse. The plasma channel modifies the index of refraction which, in the absence of relativistic effects, is given by

$$\eta_R \simeq 1 - (\lambda_o/\lambda_p)^2 (n(r)/n_o)/2, \qquad (8)$$

where $n(r)$ is the electron density profile. Optical guiding can occur when the plasma density is minimum on-axis. Analysis of the wave equation for a fixed parabolic plasma density channel indicates that optical guiding of a Gaussian laser pulse occurs when the channel density depth is given by[8]

$$\Delta n = (\pi r_e r_o^2)^{-1}, \qquad (9)$$

where $\Delta n = n(r=r_o) - n(r=0)$ and $r_e = |e|^2/m_o c^2$ is the classical electron radius [see Fig. 4]. For example, the density depression necessary to guide an optical beam having a spot size of $r_o = 30$ μm is $\Delta n \simeq 10^{17}$ cm^{-3}. Fully nonlinear simulations of pulse propagation, which include the self-consistent evolution of the density channel, indicate that this mechanism is capable of overcoming diffraction for short pulse lengths and high intensities.[8]

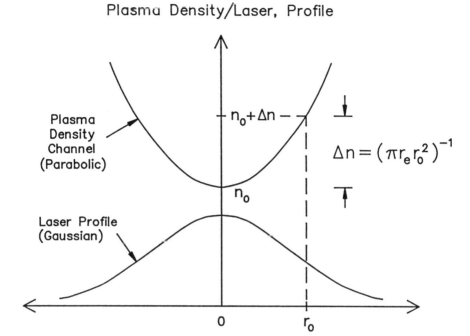

Fig. 4 Plasma density profile necessary for optical guiding in a preformed plasma channel.

The self-consistent, 2-D propagation of intense laser pulses in plasmas has been recently analyzed theoretically and numerically.[8] Figures 5a,b show the laser intensity, a^2, as a function of $\xi = z - ct$ and r. The figure shows the intensity, initially, at $z = 0$, and after propagation 10 Rayleigh lengths in a preformed plasma channel. Figure 6a shows the laser spot size after 10 Rayleigh lengths. The front of the laser pulse is optically guided while the body of the pulse focused. Figure 6b shows the evolution of the spot size at the pulse center as a function of propagation distance. The spot size oscillates about its initial value during the full propagation distance. In addition, as the pulse propagates within the plasma channel it generates a large amplitude wakefield. The accelerating gradient remains at a relatively constant value of $E_{max} \simeq 3$ GeV/m.

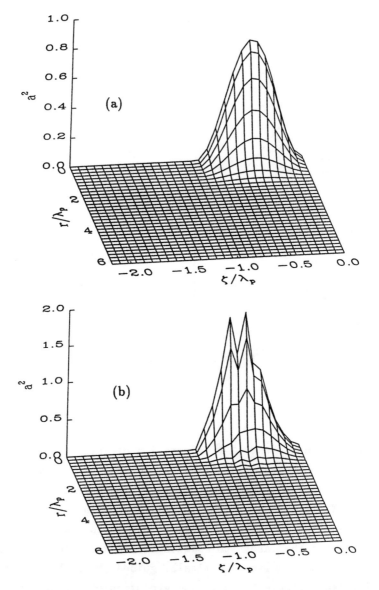

Fig. 5 The self-consistent propagation of an intense laser pulse, with $a_o = 0.9$, $\lambda_o = 1$ μm, and $c\tau_L = r_o = \lambda_p = 0.03$ cm, in a density channel with $n(r=0) = 1.2 \times 10^{16}$ cm^{-3}. (a) shows the initial intensity profile, a^2. (b) shows the a^2 profile after propagating 10 Z_R, where $Z_R = 28.3$ cm.

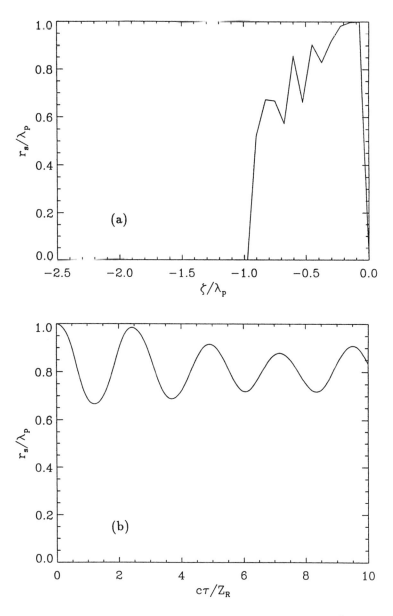

Fig. 6 The self-consistent propagation of an intense laser pulse in a plasma channel with same parameters as for Fig. 5. (a) shows the laser pulse spot size after 10 Z_R, where Z_R = 28.3 cm. (b) shows the behavior of the spot size at the pulse center as a function of propagation distance.

D. Laser Cooling of Electron Beams

The interaction of an electron beam with a counterstreaming intense laser field may be used to produce synchrotron radiation and to radiatively damp beam emittance[13] in much the same manner as a static undulator (or wiggler) magnet in a storage ring is used to generate synchrotron radiation and radiatively damp beam emittance [see Fig. 7]. An advantage in using an intense laser pulse over a static undulator is that due to the short wavelength ($\lambda_o \simeq 1$ μm) and high intensity ($a_o > 1$) of the laser pulse, extremely short-wavelength, intense synchrotron radiation may be produced and radiative damping may occur on rapid time scales (~ picosecond).

When an electron beam interacts with a counterstreaming intense laser pulse, radiation is generated by nonlinear Thomson scattering. For low intensity lasers ($a_o^2 \ll 1$), the frequency of the scattered radiation[16] from an electron beam is given by $\omega \simeq \gamma_o^2 (1 + \beta_o)^2 \omega_o$, where $\gamma_o = (1 - \beta_o^2)^{-1/2}$, $\beta_o = v_o/c$, and v_o is the initial velocity of the electron beam. In the ultra-intense regime, $a_o \gg 1$, the scattered radiation will contain a continuum of harmonic radiation out to the harmonic number N_{max}. Furthermore, the scattered radiation is well collimated within a cone at angle $\theta \sim 1/\gamma_o$ in the direction of the electron beam. The total power in the scattered radiation may be calculated from the relativistic form of Larmor's formula. The ratio of scattered power to incident laser power for a circularly polarized laser of pulse length τ_L interacting with a plasma of density n_o is given by[13]

$$P/P_o = (8\pi/3) n_o r_e^2 c \tau_L \gamma_o^2 (1 + \beta_o)^2, \qquad (10)$$

where r_e is the classical electron radius. As an example, consider a circularly polarized incident laser with $\lambda_o = 1$ μm, $\tau_L = 1$ ps, and $P_o = 10$ TW, interacting with an electron beam with $\gamma_o = 200$ and $n_o = 10^{14}$ cm^{-3} (0.48 MA/cm^2). The total scattered power is $P/P_o \simeq 3.2 \times 10^{-7}$; hence, $P = 3.2$ MW. The scattered radiation will have wavelengths in the range $\lambda \simeq \lambda_o/\gamma_o^2(1 + \beta_o)^2 \simeq 0.06$ Å. The pulse length of the scattered radiation is determined by the laser-electron interaction time, typically on the order of picoseconds.

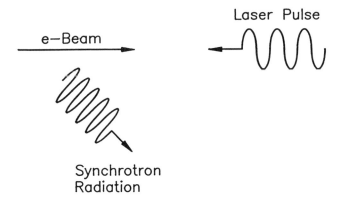

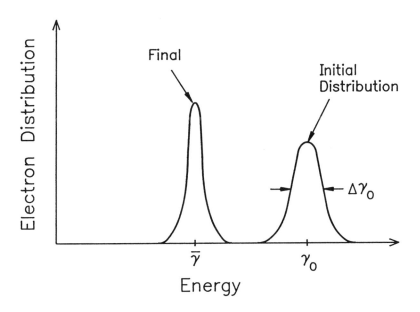

Fig. 7 Schematic of an intense laser interacting with a counter-streaming electron beam producing synchrotron radiation which leads to radiative cooling of the electron beam.

As an electron beam emits incoherent radiation via nonlinear Thomson scattering, it is subsequently "cooled", i.e., the normalized emittance and energy spread of the electron beam is damped.[13] This is the result of the radiation reaction force[17] which an electron experiences while emitting radiation. The normalized electron beam emittance ε_n will be damped according to the relation[13]

$$\varepsilon_n = \varepsilon_{no}/(1 + z/L_R), \qquad (11)$$

where ε_{no} is the initial normalized emittance, and L_R is the characteristic damping distance,

$$L_R[\mu m] = 3.4 \times 10^6 \lambda_o^2[\mu m]/\gamma_o a_o^2, \qquad (12)$$

where a circularly polarized laser has been assumed. The mean energy of the electron beam also decreases, $\bar{\gamma} = \gamma_o/(1 + z/L_R)$. Furthermore, it can be shown that the fractional beam energy spread decreases, $\Delta\gamma/\bar{\gamma} = (\Delta\gamma_o/\gamma_o)/(1 + z/L_R)$, where $\Delta\gamma_o$ is the initial beam energy spread. This radiative cooling effect can be significant. As an example, consider a $\lambda_o = 1$ μm laser with $a_o = 10$ interacting with an electron beam of $\gamma_o = 200$. The damping length for these parameters is $L_R \simeq 170$ μm. The energy lost by the electron beam appears in the form of synchrotron radiation.

III. CONCLUSION

An attempt has been made to briefly discuss how the development of compact ultra-intense lasers[1,2] may impact research in the areas of advanced ultra-high gradient accelerators, advanced synchrotron radiation sources, and radiative cooling of electron beams. An important measure which characterizes ultra-intense laser interaction physics is the laser strength parameter, a_o, which is proportional to the square root of the laser intensity. When this parameter is significantly greater than unity the electron dynamics becomes highly relativistic and nonlinear, thus resulting in a wide variety of new phenomena.[3-13] This high laser field regime has not been fully analyzed. In addition to laser wakefield acceleration,[3-5] optical guiding,[5-8] nonlinear Thomson scattering,[13] and laser cooling of electron beams,[13] many other ultra-intense laser interaction phenomena may occur, including relativistic harmonic generation,[5,11] laser frequency amplification using ionization fronts or plasma waves,[9,10] and stimulated backscattered harmonic radiation.[12] It should be emphasized that this is only a partial list of phenomena in the rapidly growing research areas of ultra-high field physics.

ACKNOWLEDGMENTS

The authors acknowledge useful discussions with J. Krall, G. Joyce, A. Ting, D. Umstadter, and G. Mourou. This work was supported by the Department of Energy and Office of Naval Research.

REFERENCES

1. D. Strickland and G. Mourou, Opt. Commun. 56, 216 (1985); P. Maine, D. Strickland, P. Bado, M. Pessot, and G. Mourou, IEEE J. Quantum Electron. QE-24, 398 (1988); M. Pessot, J. A. Squire, G. A. Mourou, and D. J. Harter, Opt. Lett. 14, 797 (1989); M. Ferray, L. A. Lompre, O. Gobert, A. L'Huillier, G. Mainfray, C. Manus, and A. Sanchez, Opt. Commun. 75, 278 (1990); C. Sauteret, D. Husson, G. Thiell, S. Seznec, S. Gary, A. Migus, and G. Mourou, Opt. Lett. 16, 238 (1991); J. Squirer, F. Salin, G. Mourou, and D. Harter, Opt. Lett. 16, 324 (1991).
2. M. D. Perry, F. G. Patterson, and J. Weston, Opt. Lett. 15, 1400 (1990); F. G. Patterson, R. Gonzales, and M. Perry, Opt. Lett. 16, 1107 (1991); F. G. Patterson, and M. Perry, J. Opt. Soc. Am. B 8, 2384 (1991).
3. T. Tajima and J. M. Dawson, Phys. Rev. Lett. 43, 267 (1979); L. M. Gorbunov and V. I. Kirsanov, Zh. Eksp. Teor. Fiz. 93, 509 (1987) [Sov. Phys. JETP 66, 290 (1987)]; V. N. Tsytovich, U. DeAngelis, and R. Bingham, Comments Plasma Phys. Controlled Fusion 12, 249 (1989); V. I. Berezhiani and I. G. Murusidze, Phys. Lett. A 148, 338 (1990); T. C. Katsouleas, W. B. Mori, J. M. Dawson, and S. Wilks, in SPIE Conf. Proc. 1229, ed. by E. M. Campell (SPIE, Bellingham, WA, 1990), p. 98.
4. P. Sprangle, E. Esarey, A. Ting, and G. Joyce, Appl. Phys. Lett. 53, 2146 (1988); E. Esarey, A. Ting, P. Sprangle, and G. Joyce, Comments Plasma Phys. Controlled Fusion 12, 191 (1989).
5. P. Sprangle, E. Esarey, and A. Ting, Phys. Rev. Lett. 64, 2011 (1990); Phys. Rev. A 41, 4463 (1990); A. Ting, E. Esarey, and P. Sprangle, Phys. Fluids B 2, 1390 (1990).
6. C. Max, J. Arons and A. B. Langdon, Phys. Rev. Lett. 33, 209 (1974); G. Schmidt and W. Horton, Comments Plasma Phys. Controlled Fusion 9, 85 (1985); G. Z. Sun, E. Ott, Y. C. Lee, and P. Guzdar, Phys. Fluids 30, 526 (1987); W. B. Mori, C. Joshi, J. M. Dawson, D. W. Forslund, and I. M. Kindel, Phys. Rev. Lett. 60, 1298 (1988); P. Gibbon and A. R. Bell, Phys. Rev. Lett. 61, 1599 (1988); C. J. McKinstrie and D. A. Russell, Phys. Rev. Lett. 61, 2929 (1988); T. Kurki-Suonio, P. J.

Morrison, and T. Tajima, Phys. Rev. A $\underline{40}$, 3230 (1989); A. B. Borisov, A. V. Borovskiy, V. V. Korobkin, A. M. Prokhorov, C. K. Rhodes, and O. B. Shiryaev, Phys. Rev. Lett. $\underline{65}$, 1753 (1990).
7. P. Sprangle, C. M. Tang, and E. Esarey, IEEE Trans. Plasma Sci. $\underline{PS-15}$, 145 (1987); E. Esarey, A. Ting, and P. Sprangle, Appl. Phys. Lett. $\underline{53}$, 1266 (1988); E. Esarey and A. Ting, Phys. Rev. Lett. $\underline{65}$, 1961 (1990); P. Sprangle, A. Zigler, and E. Esarey, Appl. Phys. Lett. $\underline{58}$, 346 (1991).
8. P. Sprangle, E. Esarey, J. Krall, and G. Joyce, to be published.
9. S. C. Wilks, J. M. Dawson, and W. B. Mori, Phys. Rev. Lett. $\underline{61}$, 337 (1988); W. M. Wood, G. Focht, and M. C. Downer, Opt. Lett. $\underline{13}$, 984 (1988); S. C. Wilks, J. M. Dawson, W. B. Mori, T. Katsouleas, and M. E. Jones, Phys. Rev. Lett. $\underline{62}$, 2600 (1989); H. C. Kapteyn, and M. M. Murnane, J. Opt. Soc. Am. B $\underline{8}$, 1657 (1991); W. Mori, Phys. Rev. A $\underline{44}$, 5118 (1991).
10. E. Esarey, A. Ting, and P. Sprangle, Phys. Rev. A $\underline{42}$, 3526 (1990); E. Esarey, G. Joyce, and P. Sprangle, Phys. Rev. A $\underline{44}$, 3908 (1991).
11. D. Umstadter, X. Liu, J. S. Coe, C. Y. Chien, E. Esarey, and P. Sprangle, in "Coherent Short Wavelength Radiation: Generation and Application," ed. by P. Bucksbaum and N. Ceglio. (Opt. Soc. Amer., Washington, DC 1991).
12. P. Sprangle and E. Esarey, Phys. Rev. Lett. $\underline{67}$, 2021 (1991); E. Esarey and P. Sprangle, accepted by Phys. Rev. A.
13. P. Sprangle and E. Esarey, to be published.
14. Efforts are underway at Limeil, France, and Osaka, Japan.
15. T. S. Luk, A. McPherson, G. Gibson, K. Boyer, and C. K. Rhodes, Opt. Lett. $\underline{14}$, 1113 (1989); S. Watanabe, A. Endoh, M. Watanabe, H. Sarukura, and K. Hata, J. Opt. Soc. Amer. B $\underline{6}$, 1870 (1989).
16. E. S. Sarachik and G. T. Schappert, Phys. Rev. D $\underline{1}$, 2738 (1970).
17. J. D. Jackson, "Classical Electrodynamics", (Wiley, New York, 1975), Chap. 17.

On possibilities of fast cooling of heavy particle beams

Ya. S. Derbenev
Randall Laboratory of Physics, University of Michigan
Ann Arbor, Michigan 48109-1120 USA

ABSTRACT

Two methods of fast cooling of intensive beams are described. The first one, coherent electron cooling, is based on enhancement of friction effect in the electron cooling method using a microwave instability of electron beam specially arranged in the cooling section. This method is effective for cooling of high-temperature circulating beams. The second one, self-cooling, is based on use of the intrabeam Coulomb scattering of particles during the adiabatic processes of beam acceleration and transverse compression. This method allows frequent decrease emittance of an intensive beam issued by a low-temperature source.

1. Coherent electron cooling

Three methods of cooling of circulating heavy particle beams are known today: electron cooling, which is based on the use of a co-moving electron beam, straight or circulating, as in a thermostat[1-3] ; stochastic cooling based on the use of RF feedback system[4,5] ; laser cooling of ion beams[6] . Electron and laser cooling are efficient to cool low-temperature intensive beam; oppositely, stochastic cooling is effective to cool high-temperature, low-intensity beams. The coherent electron cooling[7,8] described below combines principles and advantages of both electron and stochastic cooling.

First, let us give the qualitative consideration in favor of the principle possibility to increase the friction effect at Coulomb interaction. As it is known, a fast charged particle in plasma is effected by the friction force

$$\vec{F}(\vec{v}) = -\frac{4\pi Z^2 e^4 n_e}{mv^2}L(v)\frac{\vec{v}}{v} \equiv -\frac{(Ze)^2}{\rho_{sh}^2}L(v)\frac{\vec{v}}{v} \tag{1}$$

where Ze is a particle charge, e, m and n_e are respectively the charge, mass and density of electrons, $L(v) = \ell n(\rho_{max}/\rho_{min})$ is the Coulomb logarithm. The fast condition means that the particle velocity v is large compared to the electron heat velocity $v_e = \sqrt{T_e/m}$. In this case the maximum impact parameter in Coulomb logarithm is equal to the distance of dynamical shielding for effecting Coulomb forces: $\rho_{max} = \rho_{sh} = v/\omega_e$ where $\omega_e = (4\pi n_e e^2/m)^{1/2}$ is an electron plasma frequency. Since the particle is fast, this distance is large compared to the

Debay radius of electron gas: $\rho_{sh} \gg r_D = v_e/\omega_e$ which means that the plasma collective degrees of freedom corresponding to excitation in a characteristic time ω_e^{-1} of a group electron motion in the range of a radius $\sim \rho_{sh}$ are effectively participate in interaction with fast particle. This excitation, however, is so small that the contribution of collective modes into the total response is reduced to an insignificant increase in the Coulomb logarithm ($\rho_{min} \sim Ze^2/mv^2$); effectively, a fast particle interacts with its image on a distance about ρ_{sh}.

The collective response could be increased proportionally to the number of electrons in the interaction region if the initial excitations could increase spontaneously. For this, the electron plasma should be able to self-bunching, i.e. should be unstable in the region of the wave-lengths $\lambda > r_D$.

A principle amplification can be naturally inserted into the scheme of the electron cooling method. On the cooling section such conditions should be arranged that the moving "electron plasma" should become unstable in the given range of the wave lengths. Then, an excitation caused by an input ion will be transferred by electron flux developing exponentially independent of the ion; at the output from electron beam the ion will acquire the momentum correlated with its input velocity. In any case, quite a strong correlation is possible unless the excitation reaches the nonlinear regime, i.e. the density modulation within the required scale of distances remains relatively small. It is, of course, necessary to provide the optimum output phase relations in the position, and velocity of an ion with respect to electron "avalanche" produced by the ion. Such a task is facilitated by the motion of ions and electrons in the fields given is absolutely different. In particular, after interaction at the "input" the beams can be separated and then they can be made interacting again at the "output".

The mechanism of instability with the properties required can be precluded if in the cooling section the transverse oscillating with period λ_0 along the beam path magnetic field of relatively small amplitude is introduced into the longitudinal magnetic field accompanying an electron flux with Larmour period λ_L.

In such a system, which is called an undulator, three type of instabilities (a, b, c) are possible[9-12]:

a. An instability of the negative longitudinal mass can be realized in the region $\lambda_L < \lambda_0$ when an average velocity becomes a decreasing function of an energy because of an increase in the forced transverse velocity.

This kind of instability can be easily arranged at relativistic energies.

b. Near the point $\lambda_L = \lambda_0$ another kind of Coulomb instability occurs which can be called the cyclotron-ondulator instability. The use of this instability seems most preferable for non-relativistic case.

c. In relativistic region the mechanism of radiative instability can also be effective which is connected to generation of coherent radiation. The systems based on this principle acquired the name "free electron lasers" (FEL). Because that radiative instability has a narrow spectrum it is not so universal for its use in electron cooling as the Coulomb instabilities, but possibly under conditions the use of radiative mechanism can be profitable.

An additional kind of microwave Coulomb instability which seems easy to realize at non-relativistic energies is the parametric instability of longitudinal plasmas oscillation of the electron beam; it occurs when the plasma parameter of the electron beam, ω_e modulated with the frequency $\omega \approx 2\omega_e$. Such a modulation can be realized via modulation of electron beam size or velocity.

Note that the possibilities to provide the correct phase relations and finally the cooling effect are extended with the use of electron plasma oscillations in the input, output, and amplification sections.

In the linear approximation effective friction force increases by a factor $k = e^{\Lambda L_c}$, with respect to (1), where Λ is the increment of an instability and L_c is the cooling section length. There are, of course, a few limitations on k of a different kind.

An admissible value of k is limited by microwave Schottky-noise of electron beam, which also will increase because of the instability. With "normal" level of noises on the input k should not exceed the mass relation M/m, otherwise the heating dominates damping. There are possibilities for suppressing the Schottky-noise effect such as a parametric damping or collision relaxation[7,8]; as the result, k_{adm} can reach the maximum value of order temperature ratio T_i/T_{ell}. Note, that the ion temperature can reach a value about $10^7 - 10^{9°}$K, while the value of electron beam longitudinal temperature is about $e^2 n_e^{1/3} \sim 1°K$ [2,3].

An achievable value of k is limited by non-linear saturation of an instability. Taking into account electron and ion beam density fluctuations at the input, we have a limitation on k as follows:

$$k_{max} \sim \min\left\{n_e R_{eff}^3;\ \Gamma\sqrt{n_e R_{eff}^3};\ n_e\sqrt{R_{eff}^3/n_i}\right\}, \qquad (2)$$

where $R_{eff} \sim \rho_{sh} = \frac{v}{\omega_e}$, Γ is a coefficient of suppression of Schottky-noise amplitude in electron beam and n_i is ion concentration (all the parameters in (1) are related to co-moving frame, and we assume an isotropic amplification). Note that Γ cannot exceed the value $\Gamma_{max} \sim \sqrt{e^2 n_e R_{eff}^2/T_{ell}}$.

Finally, cooling time is limited by the time of kinematic correlation between particles which arise in a volume of effective interaction:

$$\tau_{min} \sim \frac{1}{\gamma} \frac{n_i R_{eff}^3}{2\pi \Delta f_0} \cdot \frac{R_{eff}}{R}, \qquad (3)$$

where Δf_0 is the spread of particle revolution frequency, R is machine radius, and γ is beam Lorentz-factor. In view of smallnessity of R_{eff} ($R_{eff} \lesssim a$, where a is beam radius), such a limitation is substantially weaker than in stochastic cooling method, especially at relativistic energies. Considering (2) together with (3), one can find an optimal amplification.

The method considered above combines principles of electron and stochastic cooling and microwave amplification. Such an unification promises to frequently increase the cooling rate and stacking of high-temperature, intensive heavy particle beams. Certainly, for the whole understanding of new possibilities thorough theoretical study is required of all principle properties and other factors of the method.

2. The self-cooling of straight beams

In principle, the electron cooling is capable of cooling a straight low-energetic proton or ion beam within an acceptable length of cooling section; it is effective due to a very low value of the longitudinal electron temperature ($T_{\parallel}^e \sim 10^{-4}$ eV) and the transverse motion of electrons being bounded because of the magnetic field[2,3]. However, in view of the longitudinal heating of electrons by the ions of the beam under cooling, the ratio between the electron's and the ion's current densities must be not less than $T_{\perp}^i/T_{\parallel}^e$. Therefore, the use of electron cooling for the ion beams with current density about ~ 1 mA/cm^2 seems to be problematic.

Now we describe the possibility of using intrabeam scattering of particles for the transverse cooling of a beam, with corresponding heating in the longitudinal direction, during the processes of acceleration and formation of the beam in a straight line[13]. There are no external heat energy transfers from the beam; in such a situation, the total beam entropy is not decreased, but there is redistribution between the degrees of freedom of the beam.

We assume axial symmetry of a beam and describe its evolution by variables:

$$\Gamma_\perp = \pi a^2 T_\perp \quad , \quad \Gamma_\parallel = \frac{\gamma v e Z}{I}\sqrt{T_\parallel} \quad ; \quad \Gamma = \Gamma_\perp \cdot \Gamma_\parallel$$

$$T_\perp = \frac{<(\Delta\vec{p}_\perp)^2>}{2M} \quad ; \quad T_\parallel = \frac{<(\Delta p_\parallel)^2>}{M} \quad ; \quad I = \pi a^2 n v Z e, \qquad (4)$$

where $<(\Delta\vec{p}_\perp)^2>$ and $<(\Delta p_\parallel)^2>$ are transverse and longitudinal dispersion of particle momenta at a given point of space related to a frame moving with an average beam velocity of $v = (1 - 1/\gamma^2)^{1/2}$, assuming non-relativistic the transverse particle motion, a, n, and τ are beam radius, concentration and current, Ze and M are particle charge and mass. We also assume for simplicity, that there is no gradient of $T_\perp$, $T_\parallel$ and n across the beam. Apparently, we can consider $T_\perp$ and $T_\parallel$ as effective transverse and longitudinal beam temperatures (related to a given point of the beam). The value $\sqrt{\Gamma_\perp}$ is proportional to an invariant beam emittance, $\Gamma_\parallel^{-1}$ as an effective beam density in the longitudinal phase space; both of them are dynamical invariants of a beam as an ensemble of particles.

The decreasing of the transverse entropy by collisions between particles then occurs under the condition $T_\perp > T_\parallel$; this condition can be maintained for a long time by longitudinally stretching the beam during acceleration and by transversely compressing the beam. The increase of the total entropy is small when these processes are performed slow with respect to the process of temperature relaxation. Note that the adiabatic process should start from a state of $T_\perp = T_\parallel$, in order to avoid a substantial increase of the entropy during the initial stage of the transverse cooling.

In the adiabatic limit $T_\perp = T_\parallel = T$, and Γ is still constant under collisions:

$$\Gamma = \gamma T^{3/2}/n = \text{const.} \quad (5)$$

Using (5) and the definitions (4) we get the adiabatic evolution of $\Gamma_\perp$:

$$\Gamma_\perp = \Gamma_{\perp o} \cdot [(\frac{Ia}{\gamma v})/(\frac{Ia}{\gamma v})_0]^{2/3}.$$

The maximum cooling effect would occur when all the process of beam formation and acceleration is performed adiabatically. In this case, the value v_0 is related to the cathode temperature, T_c, $(T_c = Mv_0^2)$ and a_0 is the beam size at the cathode, a_c. Table 1 gives an illustration of the maximum cooling effect for a heavy particle beam assuming no beam bunching after acceleration.

Table 1	
Maximum Cooling Effect	
Initial beam parameters	
Cathode temperature, eV	0.1
Beam radius at the cathode, cm	0.5
Parameters after acceleration	
Top energy after acceleration, MeV	100
Beam radius after acceleration, cm	0.02
Self-cooling effect	
Decreasing of beam emittance, times	100
Decreasing of transverse phase space volume, times	10^4

In practice, we cannot use the adiabatic process, but must use the quasia-diabatic process, when change of Γ is not zero but small. To formulate a corresponding condition, we define two parameters:

$$\lambda_d = \frac{\mathrm{Ia}}{\gamma v}\Big/\left(\frac{\mathrm{Ia}}{\gamma v}\right)'_d \, , \quad \lambda_{st} = -\frac{T_\perp - T_\|}{(T_\perp - T_\|)'_{st}}$$

where the symbol "d" denotes change of parameters in an external field without collisions, and "st" means change under collisions. To calculate parameter λ_{st} near the equilibrium state, we use the Landay collision integral[14] at the model of Gaussian distribution in temperatures $T_\perp$ and $T_\|$, then we get

$$\lambda_{st} = \frac{5}{8}\sqrt{\frac{M}{\pi}} \cdot \frac{\gamma v}{(Ze)^4 L}\Gamma,$$

where L is the Coulomb parameter $L = \frac{1}{2}\ell n\,[\gamma T^3/4\pi n(Ze)^6]$ with an order of value about 3–5.

At $\lambda_d \gg \lambda_{st}$ we can get the equation describing Γ increase:

$$\Gamma' \approx (\lambda_{st}/3\lambda_d^2)\Gamma.$$

Solving this equation, one can establish the boundaries of stability of quasia-diabatic process and calculate non-adiabatic effects.

When accelerating an actual beam, the adiabatic condition cannot be satisfied in the region near the cathode, because the characteristic time of acceleration there is about equal to the inverse plasma parameter $\omega_p^{-1} = \sqrt{M/4\pi n(Ze)^2}$, which is small in comparison to the temperature relaxation time λ_{st}/u. With the acceleration, the longitudinal temperature goes down very fast, and one must take into account intra-beam scattering which can limit its decrease, i.e. the longitudinal entropy will increase with collisions between particles. After a distance of about a_c from the cathode, we can equalize the transverse and longitudinal temperatures by having the transverse expansion of the beam and, if necessary, by deacceleration of the beam. In this state, we obtain an intermediate energy W_0 such that $T_c \ll W_0 \ll W_{max}$, with initial (maximum) radius a_0 and initial value of relaxation parameter λ_{st}. With these parameters, we can start the quasiadiabatic cooling process. In view of the presence of the non-adiabatic stage at the beginning of the beam evolution, the self-cooling effect will be less than potentially possible as was presented in Table 1 (see Table 2 for a case $|Z| = 1$).

In addition, we should note the following conditions for the beam dynamics in focusing and accelerating:

1. Axially symmetrical electrodes and solenoidal magnetic field can be used in order to keep an intensive low energetic beam from repulsion by the space charge.
2. The current distribution at the cathode and the accelerating electric field should have axial symmetry.
3. Electric and magnetic fields have to be matched in the region of the beam injection into the solenoid, in order to avoid radial beam excitation inside the solenoid, i.e., to reach the Brillouin's beam state[15].

Table 2	
Self-cooling of an Actual Beam	
Beam current I, A	1
Beam radius at cathode a_c, cm	0.5
Cathode temperature T_c, eV	0.1
Anode voltage V_A, kV	10
Longitudinal length of expansion section cm	6
Beam radius after expansion a_0, cm	3
Initial energy of the adiabatic process W_0, keV	10
Initial relaxation length λ_0, m	0.3
Final energy W_f, MeV	100
Maximum value of solenoidal field B_f, Tesla	10
Final beam radius in the solenoid a_f, cm	0.02
Final relaxation length λ_f, m	15
Cooling effect on beam emittance, times	25
Cooling effect on beam brightness, times	600

Note that the considered method of cooling is related to the Boersch effect[15], which is longitudinal heating of an accelerated beam due to intrabeam scattering at $T_\perp \gg T_\parallel$. Our observation is that one can deeply cool a beam transversally if accelerating and compressing it adiabatically with respect to temperature relaxation process.

REFERENCES

1. G. I. Budker, At. En. (Sov.) **33**, 346 (1967).
2. Ya. S. Derbenev and A. N. Skrinsky, Sov. Phys. Rev. **1**, 165 (1981).
3. V. V. Parkhomchuk, A. N. Skrinsky, Preprint IYaF AN SSSR No. 90-102 (1990).
4. S. van der Meer, CERN/ISR-RF/72-46 (1972).
5. Ya. S. Derbenev and S. A. Kheifets, Part. Accel. **9**, 237 (1979).
6. S. Schröder, et al., Phys. Rev. Lett. **64**, 2901 (1990).
7. Ya. S. Derbenev, Proceedings of the 7th National Acc. Conference, V. 1, p. 269 (Dubna 1981).
8. Ya. S. Derbenev, UM HE 91-28, Ann Arbor, Michigan (1991).
9. A. M. Kondratenko and E. L. Saldin, Docl. of Ac. Sc. (Sov.) **249**, 843 (1979).
10. N. M. Kroll and W. A. McMillan, Phys. Rev. **A17**, 300 (1978).
11. A. V. Burov, Ya. S. Derbenev, Preprint IYaF AN SSSR No. 81-33 (1981).
12. A. S. Artamonov and N. I. Inozemtsev, Sov. J. Commun. Tech. Electronics (USA), **34**, 52 (1989).
13. Ya. S. Derbenev, UM HE 91-13, Ann Arbor, Michigan (1991).
14. L. D. Landay, Zh. Eskp. Teor. Fiz. **7**, 203 (1937).
15. J. D. Lawson, "The Physics of Charged Particle Beams" (Clarendon Press, Oxford 1988).

The Performance of the Tevatron Collider at Fermilab

Norman M. Gelfand

Fermi National Accelerator Laboratory, Batavia IL 60510 [*]

July 31, 1991

Introduction

Fermilab is a national laboratory devoted to research in high energy particle physics and is the site of the world's largest proton accelerator. The beams used in the experiments are produced by protons accelerated through a series of accelerators, the last of which is the Tevatron, which raise the energy of the protons from their rest energy of 938MeV to a final energy of 900GeV or 1.44erg.

The process begins with a H^- ion source input to a 750keV Cockcroft-Walton electrostatic accelerator. The ions are then accelerated to a kinetic energy of 200MeV in a drift tube linac. The acceleration, now of protons, continues to a kinetic energy of 8GeV, after charge exchange injection into the Booster, an alternate gradient synchrotron built from combined function magnets, .

The next stage of acceleration occurs in the Main Ring, an alternate gradient synchrotron built with conventional copper-iron magnets arranged in a separated function lattice. At an energy of 150GeV the protons in the Main Ring are extracted and injected into the Tevatron. It too is an alternate gradient synchrotron but it is constructed with superconducting magnets operating at liquid helium temperature. In the Tevatron the proton energy is raised to 900GeV.

Because the primary purpose of Fermilab is to perform research in high energy physics, the requirements of the research program determine, within of course the ability of the accelerators to satisfy those requirements, the significant parameters of the beam at Fermilab, viz the energy, intensity and emittance. From the perspective of the accelerator physicist, the performance of the accelerator may not be optimized; to an experimenter, a particular set of operating conditions may be optimal. The operating conditions of the Tevatron are set to satisfy the experimenter.

---

[*]Operated by Universities Research Association Inc. under contract with the U.S. Department of Energy

This paper will describe the actual operating performance of the Tevatron, operating as a collider, and will indicate the planned upgrades which will enhance the physics results coming from the experiments being performed at Fermilab.

From Source to Injection Into the Tevatron.

From the Source through the Linac

The acceleration process begins with a H^- source. The properties of the Fermilab H^- source are summarized in Table I.

Table I.
Properties of the Source

Current	$\approx$50mA.
Pulse Length	30μs
Repetition Rate	15Hz
Number of Ions/sec	($\approx 1.4 \cdot 10^{14}$ H^- ions/s).

Upon exiting from the source the kinetic energy of the H^- ions is raised to 750keV by a Cockcroft-Walton electrostatic accelerator. They are then transported to and injected into a drift tube linac which increases the kinetic energy to 200MeV.

At the exit from the Cockcroft-Walton, we can make our first measurement of the area of the beam in phase space. These measurements are destructive and are not routinely done; they are made only for diagnostic purposes. The area A of the beam in the phase space xx' or yy' is called an emittance ϵ. It must be recognized however that the particles are not normally confined within a well defined area in the phase space. This leads to the unfortunate situation where there are several definitions of the emittance defined in terms of σ and σ', the rms widths of the measured spatial and angular distributions. At Fermilab, we have defined the emittance, assuming that the distribution in phase space is Gaussian, in terms of the area that would include 95% of the particles. As a result we calculate the emittance as $\epsilon = 6\pi \cdot \sigma\sigma'$. Other high energy physics laboratories use other definitions for the emittance: at CERN, $\epsilon = 4\pi \cdot \sigma\sigma'$ while at the SSC, $\epsilon = \pi \cdot \sigma\sigma'$.

Table II gives the emittance of the beam at the exit from the Cockcroft-Walton at a kinetic energy of 750keV, at the entrance to the linac and at the exit from the linac, where the H^- ions have a kinetic energy of 200MeV.

Table II
Emittance of the H⁻ Ions [1]

Location	Kinetic Energy	$\epsilon(h)$ (mmmr)	$\epsilon(v)$ (mmmr)
Exit from the Cockcroft-Walton	750keV	$\approx 25\pi$	$\approx 50\pi$
Entrance to the Linac	750keV	$\approx 68\pi$	$\approx 75\pi$
Exit from the Linac	200MeV	$\approx 7.7\pi$	$\approx 8.0\pi$

There is shrinkage of the emittance in the linac but that is just an artifact of the acceleration process: the longitudinal momentum has increased while the transverse momentum is not changed. Consequently q', and the area in phase space are reduced. To better characterize what is happening to the beam we define the normalized emittance:

$$\epsilon_n = \epsilon \cdot \beta_r \cdot \gamma_r$$

where $\beta_r = v/c$ and $\gamma_r = E/m_p c^2$.

Table III gives the value of the normalized emittances of the beam from the exit of the Cockcroft-Walton to the end of the linac. The normalized emittance increases by almost a factor of two in the linac.

Table III
Normalized Emittance of the H⁻ Ions[1]

Location	Kinetic Energy	$\epsilon_n(h)$ (mmmr)	$\epsilon_n(v)$ (mmmr)
Exit from the Cockcroft-Walton	750keV	1.0π	2.0π
Entrance to the Linac	750keV	2.7π	3.0π
Exit from the Linac	200MeV	5.3π	5.5π

The Booster

Upon entering the booster a thin foil is used to strip the two electrons from the H⁻ ion yielding a bare proton. (The process is 98-99% efficient.) The protons are then captured by the magnetic field of the Booster.

The Booster is a rapid cycling (15Hz) alternate gradient synchrotron which raises the proton kinetic energy to 8GeV. The emittance of the beam in the Booster depends on the number of protons being accelerated. Table IV gives the values of the emittance at the end of the Booster cycle.

---

[1] These data are from an internal Fermilab report EXP-111 *The Beam Emittance* by Sho Ohnuma, Nov. 28, 1983.

Table IV[1]
Normalized Emittance at the Exit From the Booster

- **Kinetic Energy 8 GeV**

    N is the number of protons in 10^{10} per Booster bunch

- $\epsilon_n(h) = (4.03 + 2.03N + 0.911N^2)\pi$

- $\epsilon_n(v) = (5.86 + 1.18N + 0.766N^2)\pi$

The Main Ring

From the Booster, the 8GeV protons are transported to the Main Ring where the energy is raised to 150GeV total energy. The Main Ring, for the data that I will discuss, was operating in the collider mode which is different from the mode of operation for the fixed target experiments. When operating in the collider mode the protons are coalesced into a bunch before they are extracted from the Main Ring and injected into the Tevatron.

In the Main Ring emittance measurements are not destructive and are made with flying wires located in the accelerator. The wires measure the beam profile, and with a knowledge of the lattice parameters at the location of the wires, the emittances can be computed. Table V gives the emittances of the beam at various times in the Main Ring cycle. The data are from measurements made on the beam during the last collider running period and are not the results of special measurements made during study periods. The data were taken under a variety of beam intensities and are characteristic of the emittances actually encountered during the run.

Table V
Normalized Emittance in the Main Ring

Time in the Main Ring Cycle	$\epsilon_n(h)$ (mmmr)	$\epsilon_n(v)$ (mmmr)
Injection (8GeV)	$\approx 10\pi$	$\approx 12\pi$
Flat Top (150GeV)	$\approx 10 - 15\pi$	$\approx 10 - 15\pi$
Coalesced Beam (150GeV)	$\approx 15 - 20\pi$	$\approx 15 - 20\pi$

The Tevatron

The Accelerator

The Tevatron is a large (radius=1km, the same as the Main Ring) proton accelerator constructed from superconducting magnets. It was designed to accelerate protons to an energy of 1TeV. It is currently operating at an energy of 900GeV.

In the near future, by lowering the temperature of the superconducting magnets by ≈0.4K, we will be able to raise the energy to 1TeV .

In terms of its lattice, the Tevatron is a strong focussing, separated function accelerator. In addition to the arcs, containing the bending dipoles and the focussing quadrupoles, there are 6 long straight sections located in the lattice. Of these, one is used for injecting the beams, another is used for the r.f. accelerating cavities and a third was used as an interaction region for $p\bar{p}$ collisions.

The orbits of protons and anti-protons in the Tevatron were the same and so they could easily be made to collide. [2] The problem in building the collider was to create a source of anti-protons with high brightness [3] and to preserve the brightness of the protons and anti-protons as they are accelerated and as they pass from one accelerator to another. [4]

In the collider mode, a bunch of coalesced protons from the main ring was injected into the Tevatron and stored there at 150GeV. This process was repeated 6 times so that there were 6 bunches of protons stored in the Tevatron at 150GeV. For reasons to be discussed later the emittance of the protons was then deliberately increased.

Following the injection the 6 bunches into the Tevatron, anti-protons were removed from the anti-proton source, injected into the Main Ring, accelerated to 150GeV, coalesced into an anti-proton bunch and injected into the Tevatron. This too was repeated 6 times.

The 6 proton bunches and the 6 anti-proton bunches were then accelerated to 900GeV. The lattice was then modified to increase the probability of collisions at the interaction point containing the CDF detector. The beam was also scraped horizontally to eliminate the tails of the particle distribution and to reduce the background in the detector.

As in the Main Ring, the beam emittance measurements can be made using a flying wire system to measure the beam profile. Table VI gives the emittances of the beam at various times in the Tevatron cycle. Again the data are from measurements made on the beam during the last collider running period and are characteristic of the emittances actually encountered. The measurements at low-β are made after the beam has been scrapped to reduce background in the experimental detector.

---

[2]In the next collider run the proton and anti-proton orbits will be different due to electrostatic separators inserted into the Tevatron lattice.
[3]I will not discuss the problems associated with producing the anti-protons and creating a high brightness source.
[4]The preservation of the brightness during the transfer of the beam from one accelerator to another, becomes increasingly difficult as the energy of the beam increases.

Table VI
Normalized Emittance in the Tevatron

Time in the Tevatron Cycle	$\epsilon_n(h)$ (mmmr)	$\epsilon_n(v)$ (mmmr)
Injection (150GeV)	$\approx 15 - 20\pi$	$\approx 20 - 25\pi$
Before Acceleration (150GeV)	$\approx 18 - 25\pi$	$\approx 20 - 28\pi$
Flat Top (900GeV)	$\approx 20 - 25\pi$	$\approx 25 - 30\pi$
Low β (900GeV)	$\approx 15 - 20\pi$	$\approx 20 - 25\pi$

The energy of the collisions (the energy in the Center of Mass system) was 1.8TeV. The collision of a 900GeV proton with a proton at rest has an equivalent center of mass energy of 41GeV, so that the advantage of a collider in achieving high collision energy is obvious.

Luminosity

The reason for building the Tevatron was to provide high energy beams for the study of elementary particle collisions. We naturally wish to maximize the number of collisions in a given running period. This objective has a large part in determining the characteristics of the Tevatron beam, and therefore, it is useful to begin our discussion of the characteristics with an expression for the interaction rate $\mathcal{R}(interactions/sec)$ for any reaction:

$$\mathcal{R} = \sigma \cdot \mathcal{L}$$

where:
σ is the cross section for a given reaction (cm^2);
$\mathcal{L}$ is the flux or luminosity of the collider $((cm^{-2} \cdot sec^{-1}))$.

The cross section for a typical reaction of interest is $\approx 10^{-33} cm^2$ or $1 nb$. If we say that we can collide particles for $10^7 sec/year$ ($\approx 30\%$ efficiency) then we will get $10^{-26} \cdot \mathcal{L}$ interactions per year. If we want a sample of 10^5 events to analyze (a typical desire) from a year's run, we will need a luminosity $\mathcal{L}$ of at least $10^{31}/(cm^2 \cdot sec)$

The luminosity can be expressed in terms of the properties of the colliding proton and anti-proton bunches in the Tevatron. For each colliding bunch:

$$\mathcal{L}_{bunch} = \mathcal{F}(\sigma_l) \cdot f_r \cdot \frac{N_p \cdot N_{\bar{p}}}{\mathcal{A}}$$

where:

$$\mathcal{A} = 4\pi \cdot (\sigma_h(p)^2 + \sigma_h(\bar{p})^2)^{1/2} \cdot (\sigma_v(p)^2 + \sigma_v(\bar{p})^2)^{1/2}$$

and where:

- $\mathcal{F}$ is a function of the bunch lengths.

- f_r is the revolution frequency of the beam.

- N_p is the proton bunch intensity.
- $N_{\bar{p}}$ is the anti-proton bunch intensity.
- $\sigma_{h,v}(p),(\bar{p})$ is the rms size of the bunch at the interaction point.

The total luminosity, $\mathcal{L}$, is gotten by summing the luminosities of the colliding bunches, i.e. $\mathcal{L} = \Sigma \mathcal{L}_{bunch}$.
$\mathcal{A}$ is the effective area of overlap between the proton and anti-proton beams.

The way to increase the luminosity is clear; we increase N_p and $N_{\bar{p}}$ and decrease $\mathcal{A}$. This is in fact what is done *but only up to a point*. To explain why, I will need to make a digression into the Courant-Snyder Theory of the Alternate Gradient Synchrotron.

The equation for the transverse motion for a single particle in a synchrotron can be written as

$$q'' + k(s)^2 q = F(s,q)$$

where q is a transverse variable of the particle (horizontal or vertical) and s is a coordinate along the closed orbit of the particle. $k(s)$ is the focussing force due to the quadrupoles in the lattice. Since we have an accelerator of circumference C; $k(s+C) = k(s)$. In the case where $F(s,q) = 0.0$ i.e. a linear machine the solution to the equation of motion is

$$q(s) = a \cdot \beta(s)^{1/2} \cdot \cos(\phi(s) - \phi_0)$$

$$\phi(s) = \phi_0 + \int_0^s ds/\beta(s)$$

The function $\beta(s)$ is called the amplitude function and can be made to satisfy the equation $\beta(s+C) = \beta(s)$. Note that $q(s+C) \neq q(s)$ unless $\phi(s+C) = \phi(s) + 2n\pi$ where n is an integer. If n is an integer however, the motion is unstable. We define the tune of a beam $\nu = [\phi(s+C) - \phi(s)]/(2\pi)$. If the tune is an integer, we have a resonance and unstable motion.

There is an invariant of the motion, called the emittance, which is simply the area in the phase space, so we can use the same symbol as before;

$$\epsilon/\pi = \gamma(s) \cdot q(s)^2 + 2\alpha(s) \cdot \beta(s) \cdot qq' + \beta(s) \cdot q'^2$$

where:
$\alpha = -1/2 \cdot d\beta/ds$; and
$\gamma = (1 + \alpha^2)/\beta$.
It is easy to show that the value for a in the expression for $q(s)$ is simply $a = (\epsilon/\pi)^{1/2}$ so that

$$q(s) = \sqrt{\epsilon/\pi \cdot \beta(s)} \cdot \cos(\phi(s) - \phi_0)$$

The beam in the Tevatron is composed of many particles each with its own value of ϵ. Nonetheless it is possible to relate the size, σ, of a beam of particles at a point in the lattice, with a characteristic value for the emittance for the beam. The relation is very simple:

$$6\pi\sigma^2(s) = \epsilon \cdot \beta(s)$$

(I am ignoring the dispersion of the lattice at the location of the flying wire used to measure σ, and the momentum spread of the beam.)

To decrease $\mathcal{A} \approx \pi\sigma_x \cdot \sigma_y$ we can decrease the value of β at the interaction point and we can decrease the size of the beam by decreasing the emittance ϵ.

The smallest possible value of β at the interaction point is typically limited by the strength and aperture of the quadrupoles that focus the beam, and the length of the interaction region, i.e. the size of the detector that is designed to observe the interactions.

The other way of reducing $\mathcal{A}$ would be to reduce the emittance of the beam. Unfortunately we cannot simply reduce the emittance indefinitely because of the effect of the beam-beam forces on the luminosity.

Beam Beam Forces

The bunches of protons and anti-protons interact via electro- magnetic forces. One result of these forces is to modify the tunes of the beams. In a real accelerator, like the Tevatron, where there is motion in both transverse planes with coupling between them, we can have resonances not only when the tunes are integers but also when the tunes satisfy the following equation:

$$n_x \nu_x + n_y \nu_y = m$$

where $n_x, n_y m$ are integers. If, due to the beam-beam force, the tune is shifted to a value which is resonant, then particles will be lost.

For the anti-protons the *linear* beam-beam tune shift at each crossing is given by

$$\delta\nu \approx \frac{N_p \cdot r_p \cdot \beta}{4\pi \cdot \gamma_r \cdot \mathcal{A}_p}$$

. In this expression

- r_p is the classical radius of the proton ($1.535 10^{-18} m$);

- β is the value of the amplitude function at the crossing point.

- γ_r is the relativistic $\gamma_r = E/m_p$.

- $\mathcal{A}_p$ is the cross sectional area of the proton beam at the crossing point. ($\approx \beta \cdot \epsilon$)

If the tune shift for the anti-protons is [5] larger than the distance between the tune of the anti-proton beam and the nearest resonance then we will lose the anti-protons. At the Tevatron, the accelerator was operated with a tune located between the 2/5 resonances and the 3/7 resonances. (During the next collider run the operating tune will be moved so that it lies between the 3/5 and the 4/7 resonances). The available tune space of 0.028 limits the allowed brightness of the protons.

A further consequence of the beam-beam interaction is due to the non-linear part of the interaction. The linear part of the interaction results in the tune shift mentioned above. The non- linear part causes the emittance of the anti-proton beam to blow up. This increase in the emittance reduces the luminosity of the Tevatron collider.

We have found that the maximum integrated luminosity over the length of a store ($\approx 20hr$) is achieved when the beam-beam interactions are reduced by increasing the emittance of the protons over their emittance at injection.

Past Experience

In the past, collisions at the detector began after the protons and anti-protons had been injected in to the Tevatron, accelerated to an energy of 900GeV and the lattice modified so that the value of β at the interaction point was $\approx 0.5m$. The protons and anti-protons were typically stored and collisions observed by the experimenters for $\approx 20hr$ and then the store was intentionally ended.

The luminosity was monitored by detectors located around the interaction point. The measured luminosity showed an initially rapid fall (short lifetime) followed by a more gradual decline after about 5hr. The decrease in luminosity was not due to the decrease in the number of protons or anti-protons in the Tevatron but rather it was due to the increase in the emittance of the beams. The increase in emittance was due, in roughly equal parts we think, to intra-beam interactions, beam gas interactions in the relatively poor vacuum in the warm regions of the Tevatron, and to some electrical noise in one or more devices. The length of the poor vacuum regions will be reduced in the next collider run since the separator insertions will have much better vacuum then the previous insertions. We have not identified the sources of the electrical noise so that they are a little hard to fix.

Modifications to Improve Collider Performance.

Fermilab is currently involved in a program designed to increase the luminosity of the Tevatron collider.

---

[5]The tune shift of the protons is smaller than that of the anti-protons and is not relevant to the following discussion.

Collider Run Ia (1991-1992).

The beam-beam interactions are among the most important factors limiting the luminosity of the Tevatron. We have been running with six bunches of protons and six bunches of anti-protons in the Tevatron. During one revolution of the accelerator each bunch will collide 2 times with another bunch, for a total of 12 collisions. Each of these collisions results in a beam-beam tune shift $\delta\nu$ but only at the locations of the detectors do the collisions yield any physics. The number of collisions points, and hence the beam-beam effects, can be reduced if the proton and anti-proton beams are separated except at the interaction points where the detectors are located. We will use electro-static separators inserted into the Tevatron lattice to produce the helical orbits which will result in the separation of the proton and anti-proton orbits. We are planning two low β interaction regions in this run so that the number of places where the beam-beam interactions take place will be reduced from 12 to 2 by using the separators. This will allow us to reduce the emittance of the protons. We can reduce it by only 40% due to the limitations in the Booster and Main Ring. In addition we will be making improvements in the anti-proton source to increase the number of anti-protons in each bunch.

Together with a small reduction in β at the interaction point, we expect the nominal luminosity to increase from $1.6 \cdot 10^{30}/cm^2 sec$ to $5.7 \cdot 10^{30}/cm^2 sec$, an improvement by a factor of ≈ 3.6.

Collider Run Ib.

The next step in our improvement schedule will center on increasing the production rate of anti-protons. This will require beams of higher brightness in the Main Ring since the aperture of the Main Ring currently limits the intensity that can be accelerated. The brightness of the beam in the Main Ring is in turn limited by the brightness of the Booster beam. In the Booster the brightness is limited by the intra-beam tune shift at injection due to space charge. The space charge tune shift can be reduced by increasing the energy of the Linac beam.

We currently have underway a linac upgrade program, to be completed in 1992, which will increase the kinetic energy of the protons from 200MeV to 400MeV. This will result in a increased anti-proton stacking rate and a luminosity of $9.0 \cdot 10^{30}/cm^2 sec$, an improvement by a factor of ≈ 5.6 from our last run.

Collider Run II.

In the subsequent run we expect to run at our original design energy of 1TeV. In addition we will have installed new kickers for the injection of anti-protons and so we will be able to inject 36 bunches of protons and anti-protons. Neither the proton bunch intensities nor the emittances will differ from the previous run. At this point the luminosity will be limited by the available number of anti protons. We expect that due to the higher energy the larger number of bunches and the smaller beam-beam effects that the luminosity will rise to $1.1 \cdot 10^{31}/cm^2 sec$, an improvement by a factor of ≈ 6.9 from our last run.

Collider Run III.

To significantly increase the number of anti-protons available for the Tevatron, and thereby the luminosity, it will be necessary to replace the Main Ring by a new Main Injector. It will then be possible to optimize the intensity and emittances of the bunches to achieve the maximum luminosity. With the new injector the Tevatron collider, which was designed to have a luminosity of 10^{30}, will achieve a luminosity of $5.7 \cdot 10^{31}/cm^2 sec$, an improvement by a factor of ≈ 36 from our last run.

The improvements are summarized in Table VII.

Table VII
Projected Luminosity Improvements In the Tevatron Collider [6]

Running Period	I	Ia	II	III
	88-89	1991	1992,1994	1996
Center of Mass Energy (Gev)	1800	1800	2000	2000
Number of Bunches	6	6	36	36
Protons/bunch (10^{10})	7	7	12	33
Anti-protons/bunch (10^{10})	2.9	7.2	1.2	3.7
Total Anti-protons (10^{10})	17	43	43	130
Anti-proton Stacking Rate (10^{10}/hr)	2	4	6	17
ϵ_p (mmmr)	25π	15π	15π	30π
$\epsilon_{\bar{p}}$ (mmmr)	18π	18π	18π	22π
β^* (cm)	55	50	50	50
Luminosity ($10^{30} cm^{-2} \cdot sec^{-1}$)	1.6	5.7	11	57

With this new luminosity the Tevatron at Fermilab will be doing exciting physics experiments well into the 21st century.

Acknowledgements

It is a please to acknowledge the very helpful assistance of David Finley, Karl Koepke and Charles Schmidt in the preparation of this paper.

---

[6]S. Holmes 1991 IEEE-PAC

High Brightness and the SSC

D.A. Edwards and M.J. Syphers
SSC Laboratory, Dallas, Texas, USA

1 Introduction

One may ask why a supposedly conventional high energy synchrotron like the SSC would be of any interest at a symposium devoted to the subject of high brightness beams. The answer, of course, is that both cost and technical considerations push the SSC design toward the high brightness end of the parameter space. In this note, we try to outline the reasons why satisfying the design parameters may not be entirely trivial.

2 Parameters

Since the Reference Designs Studies of 1984, the transverse normalized emittance of the SSC has been $1 \times \pi$ mm mrad. A hybrid definition of emittance spanning the electron and proton worlds is used; that is

$$\epsilon_n = \pi \gamma \frac{v}{c} \frac{\sigma^2}{\beta} \qquad (1)$$

where the symbols have their usual meaning. With this definition, the emittance contains 39% of the beam.

Given that all laboratories have different ideas about how transverse emittance should be defined, it is sometimes difficult to translate among them. We believe that the SSC emittance goal is a factor of two less than that at Fermilab during normal running and about a factor of four less than that produced by the CERN facility.

During a one-shift experiment, we were able to come within 15% of the SSC emittance goal at Fermilab at the Main Ring peak energy in the Autumm of 1988. The main differences between that exercise and our present endeavor are the requirement to achieve somewhat better performance with a new set of accelerators

Table 1: Emittance allowance through SSC accelerator chain.

Accelerator	ϵ_n (mm-mrad)	Protons per bunch
Linac	$\leq 0.5\pi$	10^{10}
Low Energy Booster	0.6π	10^{10}
Medium Energy Booster	0.7π	10^{10}
High Energy Booster	0.8π	10^{10}
Collider	1.0π	0.75×10^{10}

(in contrast to accelerators with nearly 20 years of operating experience) on the one hand, and the recognition that high brightness is a design requirement at the beginning (which it was not at Fermilab) on the other.

The present emittance budget is summarized in Table 1 along with the number of particles per bunch. The numbers pertain to the exit conditions from each accelerator, or, in the case of the Collider, at collision energy.

3 Emittance Growth and its Control

At low energy the principal concern is space charge. In particular, injection at a kinetic energy of 600 MeV into the Low Energy Booster is associated with a Laslett tune shift of about one-half, depending on presumptions about incoming emittance and so on. A substantial simulation effort is underway to model the dilution process. It has been deemed advisable to construct the Linac enclosure with length appropriate to 1 GeV, although the Linac will be built for 600 MeV kinetic energy initially. Thus, at one of the less predictable stages in the emittance history, an investment in the direction of lowering the risk will be in place at the outset.

It is intended that space charge will not be a limitation in the Medium Energy Booster, and so at its injection energy of 12 GeV, a Laslett tune shift in the 0.1 range is felt to be reasonable. But now injection mismatches play an ever increasing role in the progression to higher energy aynchrotrons, and as cycle time become longer, processes such as gas scattering and noise sources come into play.

Injection mismatches are useful examples for illustration. In transfers from

Table 2: Transverse emittance dilution factors

Amplitude Function Mismatch:

$$\sigma^2/\sigma_o^2 = 1 + \tfrac{1}{2}|det(\Delta J)| = 1 + \tfrac{1}{2}\left(\frac{\Delta\beta/\beta_o}{\sqrt{1+\Delta\beta/\beta_o}}\right)^2$$

Dispersion Function Mismatch:

$$\sigma^2/\sigma_o^2 = 1 + \tfrac{1}{2}\left(\frac{\Delta D^2 + (\beta\Delta D' + \alpha\Delta D)^2}{\sigma_o^2}\right)\sigma_p^2$$

Injection Steering Error:

$$\sigma^2/\sigma_o^2 = 1 + \tfrac{1}{2}\left(\frac{\Delta x^2 + (\beta\Delta x' + \alpha\Delta x)^2}{\sigma_o^2}\right)$$

one ring to another, steering errors, dispersion function errors, and amplitude function errors can all lead to emittance dilution, some fraction of which can be lessened depending on the decoherence time of the beam. Table 2 summarizes the phase space dilution factors due to injection amplitude function, dispersion function, and steering errors. Three figures are appended that characterize the consequences of these relationships. The heirarchy of evils starts with steering errors, followed by dispersion function problems, and then amplitude function mismatches. It is easy to initiate a substantial emittance dilution with a steering error; it is pretty difficult to make as big a mistake with an error in the amplitude functions.

This is a good situation, and as a result we approach the detailed design stage with reasonable equanimity. Beam position monitors will need to deliver signals to the control room at the 10 μ m level, but that is well within the range of present performance. The main thing is that we plan for it at the outset.

We feel that the more difficult problems will be noise of one sort or another. The excessive emittance growth in the Tevatron came from noise sources that it took a long time to identify and correct.

4 Concluding Remarks

The parameters of the SSC call for a rather high brightness beam. It is felt that the requirements can be achieved with the present design. The fundamental goal of the SSC is to reach a luminosity of $10^{33} cm^{-2} sec^{-1}$. The SSC program holds some measures in reserve, as it must. A recent lattice modification has introduced space into the long bending arcs of the collider rings to allow for a degree of flexibility in the future. As yet, no appeal has been made to the potential benefits of damping, coalescing, and cooling. We believe that the potential benifits of these techniques should be reserved for the future.

Particle Distribution Resulting From Position Mismatch

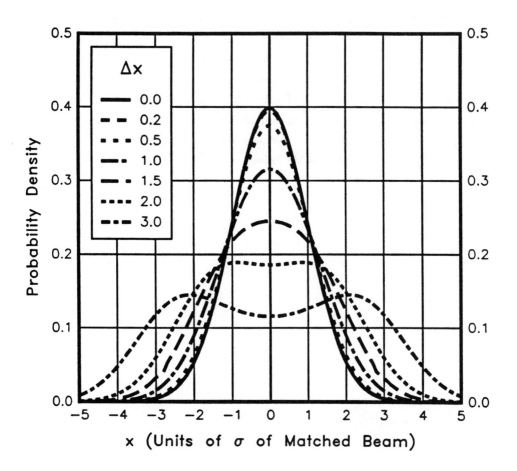

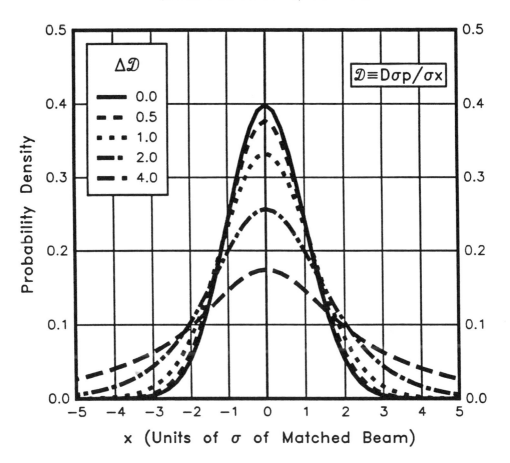

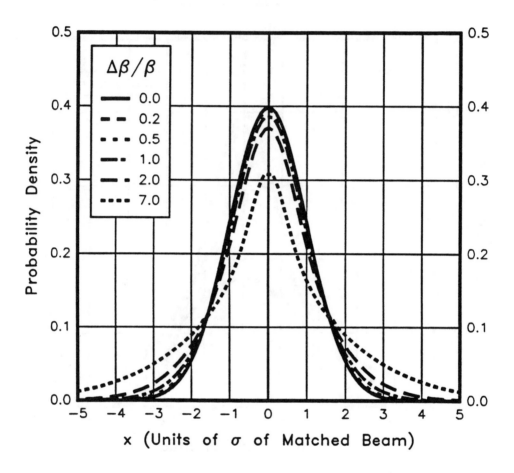

INTENSE BEAMS AT THE MICRON LEVEL FOR THE NEXT LINEAR COLLIDER[*]

John T. Seeman
Stanford Linear Accelerator Center, Stanford, CA 94309

ABSTRACT

High brightness beams with sub-micron dimensions are needed to produce a high luminosity for electron-positron collisions in the Next Linear Collider (NLC) [1]. To generate these small beam sizes, a large number of issues dealing with intense beams have to be resolved. Over the past few years many have been successfully addressed but most need experimental verification. Some of these issues are beam dynamics, emittance control, instrumentation, collimation, and beam-beam interactions. Recently, the Stanford Linear Collider (SLC) [2] has proven the viability of linear collider technology and is an excellent test facility for future linear collider studies.

PARAMETERS OF THE NEXT LINEAR COLLIDER

The luminosity of a future linear collider [3] must increase approximately as the square of the beam energy. Then, the projected event rate is constant at all energies given an expected cross section that falls with the square of the energy. The desired increase in luminosity with energy is shown in Figure 1. The methods to increase the luminosity can be seen from the equation for the luminosity L.

$$L = k\, N^-\, N^+\, f\, /\, 4\pi\, \sigma_x\, \sigma_y, \qquad (1)$$

where N^- is the average number of electrons in a bunch, N^+ is the corresponding number of positrons, k is the number of bunches in each beam, f is the repetition rate, and σ_x and σ_y are the horizontal and vertical beam sizes at the interaction point (IP), respectively. Multiple bunches, large bunch charges, high repetition rates, and small spot sizes are all desirable. All of these goals push technological limits. Multiple bunches make the task of having equal bunch energies at the end of the linac difficult as well as requiring more difficult injectors and damping rings. Large bunch charges require strong dispersion and wakefield controls indicating tight alignment and control tolerances. High repetition rates introduce considerations of reasonable AC power usages and average power issues of RF systems (modulators, loads, klystrons, ...). Small spot sizes require very small emittance sources, careful bunch length shortening after the damping rings, control of acceleration, and highly corrected final focus systems. Finally, the overall efficiency of operation must be as close to unity as possible so that the time integrated luminosity is maximized.

---

[*] Work supported by U.S. Department of Energy contract DE-AC03-76SF00515.

130   Intense Beams at the Micron Level

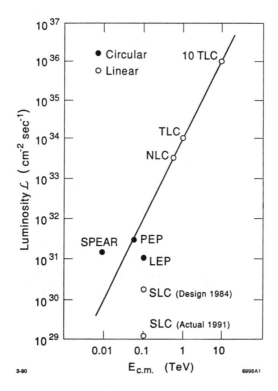

Figure 1   Desired increase in luminosity with center of mass energy for future linear colliders.

NLC REQUIRED LUMINOSITY

The required luminosity of the NLC at 250 GeV per beam is about 3×10^{33} cm^{-2} s^{-1}, which is a factor of about 33,000 over the present routine value of the SLC. The parameter changes that lead to this necessary increase are shown in Table 1. The parameters for a potential 10 TeV collider are also shown for comparison [3]. The expected increase in luminosity for the NLC comes primarily from the reduced spot sizes (X 3900) but partially from accelerating more bunches (X 10) and a higher repetition rate (X 3) but with a lower charge per bunch (X 0.28). Thus, the main task, in addition to providing the appropriate acceleration, is to increase the density of the bunches ($N / \sigma_x \sigma_y$) at the IP by a factor of 1100 from the SLC to the NLC. There are several contributing factors. The number of particles per bunch is reduced for reasons of the energy spectrum and multibunch energies. The betatron functions at the interaction point are reduced approximately by factors of 2 horizontally and 50 vertically. The increased acceleration by a factor of five reduces the emittances by that factor. At the final focus vertically flat beams are desired to reduce beam-beam effects. Thus, the injected invariant emittances provided by the sources (damping rings) must be reduced by a factor of 10 horizontally and 1000 vertically over the SLC. Consequently, considerations to preserve the vertical emittance dominate nearly all tolerance issues.

Table 1   Parameter comparison for SLC, NLC, and 10TLC

Parameter	SLC (actual)	NLC	Ratio NLC / SLC	10TLC
Luminosity (10^{31} /cm/s)	0.01	300.	X 3 X 10^4	1 X 10^5
Energy (GeV) / beam	50	250.	X 5	5000
Bunch length (mm)	1.1	0.075	X 0.07	0.015
Repetition rate (Hz)	120	360	X 3	150
Bunches / beam	1	10	X 10	120
σ_y (nanometers)	2500	4	X 625	0.1
σ_x (nanometers)	2500	400	X 6.3	27
(Charge / bunch)^2	(3 X 10^{10})^2	(1.6 X 10^{10})^2	X 0.28	(0.4X10^{10})^2

BRIGHTNESS OF BEAMS

The emittance of a particle bunch cannot be reduced to below that given by the uncertainty principle.

$$\Delta x \, \Delta p_x = \gamma \, mc \, \sigma_x \, \sigma_{x'} = \gamma \, \varepsilon_x \, mc \geq h / 2\pi \qquad (2)$$

where p is the momentum, c is the speed of light, mc^2 is the mass of the electron, γ is the relativistic energy E/mc^2, ε is the emittance, and $h/2\pi = 6.58 \times 10^{-22}$ MeV sec. A similar equation holds for the vertical emittance. It is not surprising that the invariant emittance is the natural measure of this uncertainty limit. Given the necessity of flat beams, the vertical limit is reached first. A plot of the required vertical emittance projected as a function of the beam energy is shown in Figure 2. The SLC beams are a factor of 10^8 from the uncertainty limit. However, the NLC is only a factor of 10^5 from the limit. With the present scaling laws, the uncertainty limit is reached at center of mass energies in the range 100 to 500 TeV. At those energies the flat beam requirement must be relaxed.

The beam brightness B for future linear colliders can be calculated for a single bunch given the emittances and the number of particles.

$$B = N / \pi^2 \, \gamma \varepsilon_x \, \gamma \varepsilon_y \qquad (3)$$

Examples of several accelerators with single bunch pulses (no multiple pulse stacking) are shown in Table 2 along with projections of the NLC. The achieved SLC brightness is significantly above (X 10^4) that produced by either older conventional linacs or injected and damped single bunches in a second generation storage ring (PEP). The NLC bunches must be brighter than those of the SLC by a similar factor (3 X 10^3).

132 Intense Beams at the Micron Level

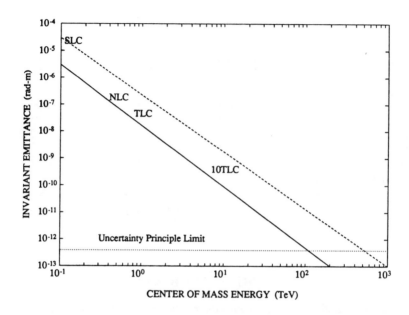

Figure 2   The uncertainty principle limit for the emittance is reached in the vertical dimension for colliders with center of mass energies of 100 to 500 TeV.

Table 2   Beam brightness for existing and future single pulse accelerators.

Accelerator	E (GeV)	N (10^{10})	$\gamma\varepsilon_x$ (10^{-5} r-m)	$\gamma\varepsilon_y$ (10^{-5} r-m)	$B=N/(\pi^2 \gamma\varepsilon_x \gamma\varepsilon_y)$ (m^{-2})
SLAC Linac*	30.	0.1	300.	300.	1.1×10^{13}
SLAC PEP**	14.5	0.2	340.	10.	6.0×10^{14}
SLC(Actual)	50.	3.0	3.0	2.7	3.8×10^{18}
NLC (Design)	250.	1.6	0.38	0.0038	1.1×10^{22}

* Before SLC upgrades.
** Single linac pulse, then damped in PEP.

ACCELERATOR PHYSICS NEEDED FOR THE NLC

There are many beam dynamics issues of concern for the next linear collider. The various subsystems of the NLC have different requirements for beam brightness and, thus, have different needs (1) for instrumentation due to phase space density, (2) for emittance control from chromatic and wakefield effects, (3) for halo collimation from backgrounds, and (4) for the beam-beam effects during collisions. Many of these issues have been resolved [4]. However, most solutions call for experimental verification.

Damping Rings

Damping rings for the NLC must operate at a high repetition rate (60 to 360 Hz) producing a horizontal equilibrium emittance an order of magnitude smaller than the SLC and with flat beams where the vertical emittance is about one percent of the horizontal. Wiggler magnets in low dispersion insertions are required to increase the radiation and thus reduce the damping time to 1 to 3 milliseconds with minimal increase in the equilibrium emittance. The choice of the optimum energy is important because of the strong effect of intrabeam coulomb scattering at low energies and strong quantum fluctuations at high energies which lead to emittance growth. Transverse and longitudinal instabilities, for example mode coupling and longitudinal multibunch effects, also influence the ring design as the bunch intensities and the number of bunches strongly effect the instability thresholds. Ring energies of 1.5 to 3 GeV are optimum in the present designs.

In addition to providing a rapid damping time, these rings must accept and damp several batches of bunches. In SLAC's NLC design ten batches of ten bunches are required. The vacuum chamber design must have a reduced impedance to avoid transverse and longitudinal instabilities. The radiation loading is not severe but requires care. To provide the required beam spacing and stability, the RF frequencies for the various designs tend to be higher than that of present rings reaching values of 1 to 3 GHz. The resulting bunch lengths are 1.3 to 8 mm, depending on the design. The use of many bunches requires feedback systems similar to the B Factory designs and common investigations can be made.

To test these systems several experimental studies are underway. There is a proposal to build a version of the damping ring for the Japan Linear Collider (JLC) [5] at KEK, which could start construction in 1992. At SLAC studies are under investigation to operate the SLC damping ring in an uncoupled regime to study the production of very flat beams and provide a decoupled beam to the Final Focus Test Beam [7].

Bunch Length Compression

In order for the main NLC linac, operating at x-band frequencies, to accelerate the bunches efficiently, the bunch length must be quite short. Lengths (σ_z) on the order of 0.03 to 0.08 mm are needed. The bunch lengths in the damping rings are, on the other hand, about a factor of 100 larger. Bunch length shortening is done with magnetics compressors. The new designs have two compression sections separated by a preaccelerator of 5 to 15 GeV to avoid the strong chromatic emittance enlargement expected in a single compressor. Each compression section contains an accelerator phased at 90 degrees to produce an RF induced head-tail energy spread of order 1 percent followed by a non-isochronous bend. The tolerances on the chromatic and geometric effects in these transport lines have a similar flavor to those calculated for the SLC Arc system. Anomalous dispersion, magnet rolls, RF phase errors, second order chromatic terms, ring coupling, magnet stability, and feedback systems are all important.

Linear Accelerator

The main NLC linear accelerator provides about 250 GeV of energy gain. The accelerator between the two bunch length compression sections contributes about 5 to 15 GeV. These accelerators must not increase significantly the invariant emittances of each bunch. (The absolute emittances, however, are reduced directly though acceleration, inversely with energy.) Furthermore, to keep within economically and environmentally acceptable limits, the total accelerator power should be limited to below 100 to 300 MW, including any future upgrade to increase the beam energy. An x-band frequency is the likely choice as it provides more gradient per stored joule of energy than present s-band structures and has parameters not too far from conventional working klystrons.

The RF power source must provide approximately 100 MW over a 100 to 800 nsec pulse. Several designs are under intense investigation. Several possible devices are nearing the power levels needed when combined with RF pulse length compression schemes. Several examples have been discussed at this symposium.

There are many effects in the linac that can increase the emittance. In the SLC the spot sizes are such that wakefield effects dominate the chromatic effects in most regions of the accelerator. In the NLC the reverse is true. However, the NLC beam intensities are at a level where transverse wakefield effects enter whenever significant accelerator parameters are adjusted.

The acceleration of multiple bunches requires that the RF structures have sufficiently small long range transverse wakefields so that trailing bunches do not experience an increase in their emittances. Therefore, the structures will include a combination of transverse waveguides to damped higher order modes and cell-to-cell detuning to dephase the higher order fields. Furthermore, the captured modes in the structure that are driven by long trains of bunches must be detuned to avoid unwanted multiple beam breakup.

The injection conditions of the beam from the bunch compressors into the linacs are important. Betatron mismatches filament along the linac increasing the emittance. The betatron functions must be matched to about 30 percent. The residual dispersion must be matched to below a millimeter compared with the 1 cm value at the SLC. Static injection errors if uncorrected cause the beam to smear producing a larger beam size and emittance. Second order contributions to the spot size, for example T_{166}, T_{266}, and T_{116} matrix elements, also lead to filamentation. Any quadrupole rolls will skew the beams and cause mixing between the planes. This mixing will produce either real or projected increases of the emittance. The rolls must be controlled to a few milliradians.

The corrected trajectory produced by steering the beam to position monitors with offset errors introduces chromatic and wakefields effects. Also contributing to these effects are the alignment errors of the quadrupoles and accelerator structures. All are further complicated by the effects of energy and energy spectrum changes during acceleration. Careful control of the lattice design and steering algorithms can reduce the effects of chromatic steering on the trajectory and the emittance [8].

The effects of dispersion, betatron mismatches, and wakefields can produce beam dimensions at the end of the accelerator with non-gaussian distributions. Special methods and controls are available to model and control these changes [9].

Feedback systems that work pulse-by-pulse are essential for keeping the beam parameters within acceptable limits for collisions. Beam positions, angles, and energy are controlled routinely in the SLC. Energy spectrum feedback is also important but has proven not to need pulse-by-pulse control. Modern control theory is used to provide cascaded control loops from the beginning to the end of the accelerator that minimally interfere with each other and thus provide maximum control. Reduction of oscillations with frequencies up to one sixth of the accelerator pulse rate can be expected.

Injection jitter that is too fast for the feedback system must be controlled by the elimination of the source or by reducing the beam's sensitivity. Controlling current fluctuations in power supplies, quadrupole transverse vibrations, RF phase and amplitude changes, noise introduced by the feedback system, and man-made seismic activity are of active concern. High repetition rates are best for feedback. The weakening of the effects of jitter on wakefields is also important for emittance control. The use of BNS damping [10] and potentially autophasing [11] are important for operation. The SLC uses BNS damping continuously with great advantage [12]. Examples of observed induced oscillations in the SLC linac with BNS damping in use are shown in Figure 3 and illustrate the complex influence of transverse wakefields on beam trajectories [2].

The halo of a beam is an enlarged population of particles at displacements of 4 to 10 sigma. The population is greater than that expected from gaussian profiles even though only a small fraction of the beam is involved. If these particles enter the final focus, they cause undesirable backgrounds in the physics detector. Removal upstream is important. The present SLC collimation design is not adequate for the next collider as the presence of any asymmetrically placed metal surfaces close to the beam, when combined with the expected small emittances, makes strong wakefields effects. The beam emittance with incorrectly designed collimators can increase significantly [13]. Additional work is needed on collimation for a fully integrated design.

The long term position stability of the accelerator and quadrupole supports and the alignment techniques of the accelerator components depend strongly on the temperature stability in the tunnel and any associated active controls. Temperature control and monitoring at the 0.5 F° level, stable support designs with reduced temperature sensitivities, and new automated alignment procedures are needed [14].

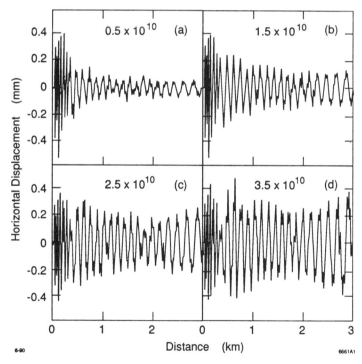

Figure 3   Observed driven single bunch oscillations along the SLC linac for various bunch charges.

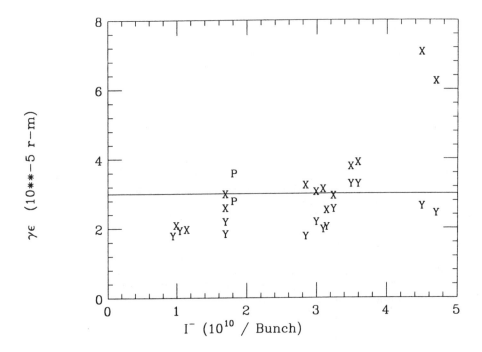

Figure 4   Measured beam emittances at 47 GeV in the SLC as a function of single beam intensity. The design emittances were reached with beam currents up to 3×10^{10} during colliding beam conditions. With the use of proven emittances controls, the emittances above 3×10^{10} are expected to be reduced in the near future.

Operation and studies of the SLC have shown that these emittance enlargement effects including chromatic and wakefields effects can be controlled during colliding beam operations to provide significant luminosity [6]. The best measured single bunch emittance conditions of the SLC to date are shown in Figure 4. These results were also obtained during collisions with three bunch colliding operation (e+,e-, e- scavenger) up to 3×10^{10} particles per bunch. The extension of these techniques to higher currents is now under active study as intensities of near 4.5×10^{10} are needed for the next run.

Finally, experimental verification of the accelerator physics effects and cures involved with the NLC are needed before a sound proposal can be made. Many experiments using the SLC to test beams with more realistic NLC parameters including flat beams and low emittances are being considered to be performed over the next several years. In addition, several other laboratories around the world are actively performing tests. Hopefully, the time and the staff needed for these vital linear collider studies will continue to be fully supported.

Final Focus

The very high density beams in the final focus of the next linear collider present many challenges. These new challenges have been dealt with, by in large, using special techniques [3], several to be tested in the Final Focus Test Beam [7] at SLAC.

The required spot sizes are small and flat. The vertical to horizontal size ratio is about 1 to 100. The chromatic correction problem in the final focus is easier with flat beams, and this advantage is fully used in the design. The small vertical size of 3 to 10 nanometers is a difficult size to measure with normal profile devices and many new devices are under consideration.

The two opposing beams are strongly disrupted because they focus each other during collision. The incoming and outgoing beams must use separate vacuum chambers. The incoming chambers have small apertures and use small bore, high gradient quadrupoles. The out going beam has a very large angular spread and needs adequate clearance for reducing backgrounds. Thus, a crossing angle is desired which requires the use of a longitudinal-transverse tilt to the beam at the interaction point to keep the beam aligned during collisions. This tilt is introduced with a transverse RF cavity near the final focussing quadrupoles, in a scheme called crab crossing [3]. On the other hand, disruption is very helpful by producing beam-beam deflections that provide a relatively non-interfering method to align the beams in collision and to determine their spot sizes. Also, the mutual focusing provides some centroid realignment if the two beams are slightly misaligned. The disruption is also expected to increase the luminosity by shrinking the beam sizes. However, to satisfy other constraints, the disruption has been reduced to a point where only minimal luminosity gain can be realized. Also, photons generated by the beam-beam forces interact with the opposing beam causing a low energy background from pair production [15]. This background can be controlled with the choice of beam parameters and the careful design of the masking of the physics detector.

Optimization techniques for reducing the spot sizes, setting the skew parameters, and correcting the timing for these small spots have been devised but need testing [16]. The Final Focus Test Beam at SLAC has been designed to test these ideas with the goal of focusing a specially prepared SLC electron beam to a 60 nm by 500 nm spot. This beam line is scheduled to be operational in late 1992 with initial tests of understanding tuning optimization, focusing schemes, small size monitoring, and stability tests for NLC style beams. The spot size as a function of the angular divergence and β^* will be studied, as well as the improved instrumentation. New alignment techniques and vibration damping useful for NLC final focus designs are also a necessary component of these tests.

A LOOK TO THE FUTURE

The NLC must be based on a design that is technically sound, operationally stable and reliable, and is economical. The technology will most likely be as conventional as possible to take advantage of proven techniques. Every new technological extension must be experimentally tested. The overall flexibility of the NLC accelerator is very important because a broad range of parameters makes the design more robust and more likely to commission rapidly. Having many options for give-and-take parameters for increasing the luminosity was very important for the SLC and will also be important for the next collider. For example, planning for up to 2 to 3 times the charge per bunch and having options to collide up to two to three times the number of bunches are reasonable. Having the potential to reduce beta functions at the IP to significantly below the design and planning for enlarged spot sizes throughout the final focus system are prudent. Background insensitive detectors are also important. Feedforward systems that fix beam errors on the same beam pulse are desirable if not necessary. Adequate commissioning time and beam monitoring are needed to provide for rapid initial progress. Issues concerning the brightness of the beams need continual attention. Finally, the accelerator should be extendable to the next level of energy operation and luminosity level without major rebuilding.

What will the global design look like? Here are my guesses. The beam energies will be about 300 GeV, extendable to 750 GeV. The length of each linac will be about 10 km. The RF frequency will be 9 to 14 GHz. The alignment tolerances will be 25 to 50 microns upstream of the IP. The transverse support movements from vibration must be kept to below 10 nm. Beam position monitors will have a resolution of about 5 microns. Pulse-by-pulse feedback systems will be globally spread over the project with many feedforward systems incorporated. Overall, the likelihood that NLC technology will experimentally converge to an acceptable level in the next 5 years is very good.

REFERENCES

1. B. Richter, *Part. Accel.* 1990. 26: 33-50.
2. J. Seeman, *Annu. Rev. Nucl. Part. Sci.* 1991. 41: 389-428.
3. R. Palmer, *Annu. Rev. Nucl. Part. Sci.* 1990. 40: 529-592.
4. D Burke, et al., "Linear Collider," 1990 DPF Summer Study on High Energy Physics, Snowmass, CO, July 1990, and SLAC-PUB-5597 (1991).
5. K. Takata and Y. Kimura, *Part. Accel.* 1990. 26: 87-96.
6. J. Seeman, et al., US PAC San Francisco 1991, and SLAC-PUB-5437 (1991).
7. D. Burke, et al., US PAC San Francisco 1991, and SLAC-PUB-5517.
8. T. Raubenheimer, SLAC-PUB-5355 (1990).
9. J. Seeman, et al., US PAC San Francisco 1991, and SLAC-PUB-5440.
10. V. Balakin, et al., Proc. 12th Intl. Conf. High Energy Accel. ,FNAL, p. 119 (1983).
11. V. Balakin, *Proc. of 1st Intl. Workshop on The Next Generation Linear Collider*, Stanford: SLAC-Report-335, p. 56 (1988).
12. J. Seeman, et al., SLAC-PUB-4968 (1991).
13. L. Merminga, et al., US PAC San Francisco 1991, and SLAC-PUB-5507.
14. G. Fischer, *Part. Accel.* 31:47-55 (1990).
15. P. Chen, US PAC San Francisco 1991, and SLAC-PUB-5557.
16. J. Irwin, US PAC San Francisco 1991, and SLAC-PUB-5539.

ADVANCED HIGH-BRIGHTNESS ION RF ACCELERATOR APPLICATIONS IN THE NUCLEAR ENERGY ARENA

R. A. Jameson
Los Alamos National Laboratory, Los Alamos, NM 87545

ABSTRACT

The capability of modern rf linear accelerators to provide intense high quality beams of protons, deuterons, or heavier ions is opening new possibilities for transmuting existing nuclear wastes, for generating electricity from readily available fuels with minimal residual wastes, for building intense neutron sources for materials research, for inertial confinement fusion using heavy ions, and for other new applications. These are briefly described, couched in a perspective of the advances in the understanding of the high brightness beams that has enabled these new programs.

DEVELOPMENT OF THE UNDERLYING BEAM DYNAMICS PRINCIPLES

It should always be appropriate, when looking ahead, to recall how we got to the present, for there are circles in circles as we progress, usually slowly, and sometimes only a little more elaborately. It is especially interesting, on this occasion of Martin Reiser's Jubilee, because fifteen years ago the prospects for using high-brightness ion beams in solving important societal problems was bright. As it is now. However, we have not yet had the chance to bring our ideas to real fruition; the challenge of our bright prospects still lies ahead and is extended to the present workers in the field, especially the younger ones. It is clear we must not only be technical experts, but must work harder to provide clear information about how accelerators could bring unique answers to the problems we would like to help solve.

Here are some of the high points, as I remember them...

In 1975-76, LAMPF, still the world's most intense ion linac, had been brought to design intensity. There was interest in using accelerator-driven systems for electronuclear breeding and intense neutron sources[1,2]. At Los Alamos, we also began to seriously consider the technical problems of using a neutral particle beam for missile defense, a program of searching for "the ultimate" in performance, with even the minimum requirement demanding about four orders of magnitude improvement in brightness over the then state-of-the-art. Heavy-ion fusion (HIF) was proposed[3,4]. Here again was an application demanding an enormous extension in brightness — 6-D brightness, because the beam bunch must not only properly overlap the target but must deposit its energy in a controlled way over a specified period.

In 1977, it was discovered[5], via simulation studies, that the parameter with the strongest influence on brightness is the rf frequency, with higher brightness possible at higher frequency (up to a point where practical limitations intrude). Lysenko quickly verified the result theoretically[6]. It was considered very surprising and sparked much discussion and lengthy investigation into the full explanation, which Wangler will outline at this meeting. Basically, there are fewer particles per bunch at higher frequency, so the space-charge forces are less, but there is a second, more subtle effect in that a smaller beam, produced by strong external focusing, is better for preserving brightness than a larger beam, contrary to intuition.

There was also very lively debate about the rms beam properties as manifested by the envelope equations for transverse and longitudinal motion, developed extensively at

CERN by Sacherer[7] and used as the design basis for their new linac then under construction. At that time, we were struggling with nomenclature and wondering about the best ways to condense the results -- the use of the terms "phase advance" and "tune depression" were very new to linac people! The discussion got severe when, in HIF studies trying to determine how much power could be transported through a focusing channel, one analysis had beam emittance in the numerator and another analysis had it in the denominator! It was Prof. Reiser who cleared it up[8], and in so doing emphasized a very important lesson -- that a new problem requires a completely fresh outlook, and that the best solution will be determined as much by the constraints imposed and the choice of fixed or variable parameters, as by the basic principles. In other words, you have to ask the right questions. I used to recommend this article of Reiser's as part of the necessary "Art of War"[9] of the accelerator designer, and I still do.

A "Space Charge in Linear Accelerators" Workshop[10] was convened at Los Alamos in the fall of 1977 (in those days, having a "workshop" was also a new idea - hard to imagine now...) to discuss experimental, calculational, and theoretical aspects of space charge phenomena and to propose future work. There was plenty of experimental evidence of unexplained phenomena, fierce arguments about the definition of emittance, the notion that there must be a stable "equilibrium" "matched" distribution, and a very strong needle from the theoreticians that "Theory has to approach reality, of course, but I think it's a two-way street; that is, I think reality has to approach the theory and I mean you really have to shape up!" (Sho Ohnuma). This was to suggest that even the basic power of the rms equations and certain aspects of KV distribution analysis had never been experimentally verified (largely because constructed machines quickly went operational and there was little time for machine physics studies), that the theory was really ahead of the practice, and that dedicated accelerator research programs were needed. This certainly fit the notions at Los Alamos, where the new Accelerator Technology Division was being formed, of LBL for HIF studies, and of Prof. Reiser, who began to implement a meticulous experimental program at the University of Maryland that continues to the present.

The Space Charge Workshop was also instrumental in a breakthrough -- the RFQ accelerator. J.J. Manca was asked to speak about new accelerator structures, and discussed work in the Soviet Union on a low-beta structure that handled large current (>100 mA) at acceleration rates of ~1 MeV/m, with injection possible at 50 keV, a capture coefficient close to 100%, and an almost constant transverse emittance. This had to be listened to seriously; the next day, the translation of his sole reference was carefully checked, a library search produced a few more papers, and we were on our way. A new application, an intense neutron source for developing fusion materials (FMIT)[11] using a 35 MeV, 100 mA cw deuteron beam on a flowing, molten lithium target, provided the initial development funds.

In late 1978, the rms approach got a severe challenge, because the observed emittance of the new CERN linac far exceeded predictions. It was found that the effects of mismatch and/or mis-steering on beam emittance had not been considered; these effects were included in the Los Alamos simulation codes, which, when applied to the experimental data, were able to bracket the observed results[12]. This restored the confidence we badly needed to pursue our on-going high-brightness programs. There was, however, another underlying component of the emittance growth observed on that linac and on all the earlier ones at LAMPF, BNL and FNAL. It had long been conjectured[13] that some sort of "thermodynamic-like" energy balance between the degrees of freedom would influence emittance growth, but the way to characterize and quantify the physics was elusive. In 1981, it was discovered[14] that a very simple local rms equipartitioning requirement on a bunched, matched injected beam would result in remarkably small emittance growth, at least in a full-scale, nonlinear simulation of a linac. As outlined more completely in the reference, one starts with the two envelope equations that require transverse and longitudinal matching:

$$\varepsilon^t = \sigma^t a^2/N\beta\lambda \text{ and } \varepsilon^l = \sigma^l b^2/N\beta\lambda,$$

(relating emittance ($\varepsilon^t, \varepsilon^l$), phase advances ($\sigma^t, \sigma^l$) including linear space-charge forces over the transverse focusing period ($N\beta\lambda$), and the transverse and longitudinal beam radii a and b), and solves them simultaneously with a third "equipartitioning equation":

$$\varepsilon^l/\varepsilon^t = \sigma^t/\sigma^l = b/a$$

that results from the requirement that the average energy in each of the coupled degrees of freedom be equal. Figure 1 shows the first computed result; the matched and equipartitioned uniform beam has small emittance growth over a 72 cell, constant accelerating gradient linac in which the quad strengths were adjusted to hold the transverse tune depression constant at 0.3. The initial longitudinal space charge parameter μ^l was ~0.8. The $\varepsilon^t/\varepsilon^l$ ratios given in the figure are the initial conditions at injection into the linac (= 0.96 for the equipartitioned case); when the energy partitioning was unbalanced via an emittance ratio of 1.5 in either direction, a clear energy transfer occurs.

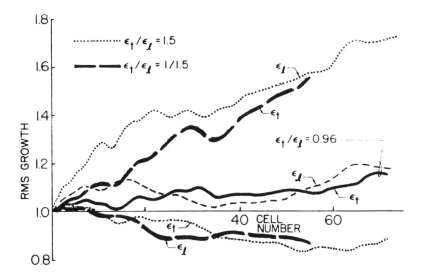

Figure 1. For the strongly beam-current loaded linac described in the text, rms emittance growth is very small when the injected beam is both matched and equipartitioned, but energy transfer occurs if the equipartitioning is not satisfied (by an rms emittance unbalance of 1.5:1 in either direction) and rms emittance growth occurs in the plane of initially lower energy, with a reduction in the rms emittance of the initially higher energy plane.

This result is easy to use and made the practical design goals clear: require local matching and energy balance and a uniform beam distribution (although Sacherer had shown that the rms properties do not depend strongly on the particle distribution). As Gluckstern pointed out at this Symposium, "Emittance growth is not inevitable". These are still goals, because practical linacs require transitions from one type of structure to another, and are subject to a host of constraints that can conflict with the matching plus equipartitioning requirements.

The simultaneous solution of these simple equations and others derived from the rms theory provide a great deal of insight and afford conceptual high-brightness designs to be quickly evaluated. I am presently building an accelerator system modeling and costing code[15] using the rms theory as the accelerator technical model. RFQs, several kinds of drift-tube linacs, and high-beta coupled-cavity linacs are included. The designer sets up a number of equations for desired properties corresponding to the number of variables he has left free, and solves them, subject to constraints, yielding structure and beam related information. For cases in which parameter changes are not too abrupt and where emittance growth is reasonable small, the addition of Liouville's theorem of emittance conservation under adiabatic acceleration as a specific equation yields results that agree well with simulation studies. It is planned to add Wangler's emittance growth formulas as a next step.

The search for more of the underlying physics of emittance behavior continued, with more breakthroughs in 1984-5 from Struckmeier, Reiser and Wangler[16], showing how non-uniform field energy in the particle distribution caused rapid emittance growth as the distribution moved to equilibrium, and showing the asymptotic behavior of the equipartitioning process. This work continues (see Wangler's paper this meeting) to try to analytically quantify further the effects of mismatch, mis-steering, identify growth rates, and very importantly, to unlock the secrets of why a diffuse halo forms around the central beam core.

This halo generation, still almost completely not understood, is at the heart of the specification of high-intensity linacs for factory-type applications where very low beam loss is desired so that "hands-on" maintenance can be performed over the facility lifetime. We know that very careful control of the beam rms properties and a uniform distribution are necessary for proper halo control, though not sufficient. So for now, we augment the rms design procedure with engineering "stay-clear" factors to keep the ratios of transverse aperture to rms beam radius, and longitudinal phase acceptance to bunch rms phase length, large.

These principles have largely been demonstrated experimentally now to the physics level, if not to the full engineering and application level that we still hope for. The RFQ accelerator, now extensively developed, is the primary demonstration, in a very fundamental way, of all our present design knowledge, as the beam is guided from dc to bunched by a complex orchestration of the parameters. The elegant experiments of the LBL HIF program and at U. Md. have consistently verified the theory. The Los Alamos SDI program has demonstrated the basic principles of funneling (where two beams are interlaced onto the same path with predicted emittance behavior) and also high-brightness performance in Alvarez linacs. This work has certainly fulfilled Ohnuma's observation that good experiments would provide the confirming evidence.

— The Accelerator Requirement for Tasks At Hand

A perspective can be placed on our confidence in the present state-of-the-art concerning high-brightness ion rf linac physics and engineering (not discussed here at all) practice by observing that today's application proposals, to be discussed below, are in the class of providing 1.5 GeV proton beams at currents up to 250 mA cw, with transverse rms normalized emittance of less than 0.1 pi-cm-mrad (computed for non-ideal beams) and with low-enough beam losses to afford hands-on maintenance[17].

We have followed through on the design principles above by using frequencies as high as possible, requiring extra-large aperture-to-beam ratios (about 20 in transverse, compared to about 6.3 at LAMPF), no permanent magnet quadrupoles (to avoid long-term radiation damage), a transition to the high-beta linac at as low an energy as possible with emittance filtering as an option at that point, and strong transverse focusing (zero-current phase advance of at least 70 degrees per period). The resulting point-design has two 350 MHz, 125 mA RFQ/DTL legs, funneled at 20 MeV into a 700 MHz DTL to 40 MeV, where emittance filtering might occur, and then acceleration in a 700 MHz, 250 mA coupled-cavity linac (CCL) to 1600 MeV. One notes that the transition energy at 40 MeV is unusually low. The CCL is configured in 7 sections of n-cell accelerating cavities, where n=2 in the first section and =10 in the last. Each accelerating cavity is accompanied by a quadrupole and a beam diagnostic device, thus forming a transverse focusing lattice unit. The short units allow strong transverse focusing to be applied.

The most important scaling factor is the number of particles per bunch. At 250 mA cw, we need only a factor of 4 more than LAMPF operates with, and even then the beam dynamics conditions are not greatly different. The factor of 250 in average current above LAMPF comes from duty-factor increase to cw, filling all the rf buckets, and the x4 increase in particles per bunch. We have carefully calibrated the beam-loss performance of LAMPF to our simulation codes, which predict the location and approximate magnitude of LAMPF losses, and made correlations to our 250 mA cw design. We need a fractional beam loss about 10 times lower than LAMPF's typical value ($<2 \times 10^{-7}$/m) for most of the CCL; the large aperture ratios of our design will insure this, and also, a cw machine has about a factor of 2-3 advantage because it does not have the leading edge beam losses associated with pulsing.

The most important challenge, to make the applications more economically attractive, is not the accelerator itself at all, but the rf power system. The direct power requirement for a 1.6 GeV, 250 mA cw beam is 400 MW, and the average beam loading in the room-temperature APT point design is 77%. The cost per average rf watt is by far the most important factor in both capital and operating costs; reductions in that factor would allow the use of a higher accelerating gradient, with further savings in the accelerator as well. Preliminary study of the cost scaling factors indicates that frequency is not very important over the range of interest, but that a doubling of the amplifier size will reduce the cost/watt by about the square root of two. Larger, and thus fewer, amplifiers may also be more reliable. Little R&D in high-power rf tubes has occurred in the past two decades except in the USSR. The magnicon amplifier[18] being developed at Novosibirsk by Nezhevenko seems like a good candidate. He proposes a 1.4 GHz cw 4 MW tube with a beam voltage of 200 kV, and gain of 45 dB. The amplifier can operate drive-modulated, and has better phase and amplitude sensitivity to drive and beam voltage than a klystron. The amplitron does not require a bunched beam, and thus is less prone to oscillations from reflected electrons and is naturally suited to multiple outputs of, say, 1 MW each, with inherent phase stability between outputs.

There is a strong need and challenge for accelerator technologists to turn their many tools and talents to the generation of cheaper rf power.

HIGH BRIGHTNESS ACCELERATOR APPLICATION CHALLENGES

Over the past decade, the case for particle beams in major nuclear energy arena applications has become stronger and stronger, as more people come to understand the basic properties of the particle beams themselves, and the fact that accelerators are reliable, proven devices with a long track record of development and operation[19]. However, for energy production and waste transmutation, as in the SDI arena previously, accelerator-driven approaches are considered new in relation to earlier ideas, and it takes a great deal of effort to develop and communicate the clear arguments necessary to make the case that accelerator approaches should be carried through to practical demonstration. The increasingly stringent regulatory environment makes it even harder to bring new technical approaches to fruition.

Nuclear Energy Sources

— Heavy-Ion Inertial Confinement Fusion Energy

After the start mentioned above, work on the idea of driving inertial confinement pellet implosions with heavy ions, whose behavior would be semi-ballistic in comparison with lighter ions, electrons, or photons, was pursued[20], at low funding levels, on target/reactor studies, on ion induction linac drivers at LBL[21], and on some aspects of the rf linac/storage ring driver approach at GSI where two new storage rings began esperiments last year[22]. Recent extensive reviews of ICF options capable of eventual energy production strongly recommend heavy-ion drivers as the highest priority approach[23].

As outlined by Hofmann[24], the apparent need to use indirectly driven targets requires a beam power density on target of $\sim 10^{16}$ w/g, and a smaller spot size by a factor of about 9 and a reduction in the quantity (transverse emittance x momentum spread) of about 27 below the already very stringent specifications for the directly driven targets considered earlier. This places a heavy burden on the driver brightness requirements. The driver could be an arrangement of funneled HI rf linacs driving a set of storage rings -- an obviously complex arrangement, one in which the main questions are not in the linacs but in the storage rings. The tests underway at GSI will be of great value in clarifying the storage ring issues. The indirect-drive requirement has stimulated design work on "non-Liouvillean" methods for filling the rings to higher brightness, and also would require doubling of the number of beam lines (from 20 to 40) to the target chamber. The other driver approach is the ion-induction linac being investigated at LBL. In the search for optimum technical solutions and lower economic costs, the induction linac driver has also evolved, first into a multiple-beam device, and recently into a recirculating accelerator, with attendant new challenges. There is no clear technical choice between the driver candidates at present, and both, in their present form, would produce electricity at an unattractively high cost.

— Nuclear Energy Production Using Accelerator-Driven Transmutation Technology

A very interesting new method for efficient nuclear energy production from ^{232}Th or ^{238}U without a long-term, high-level waste stream has been discovered at Los Alamos[25,26]. Alternatively, the system can be configured to emphasize nuclear waste transmutation. Earlier ideas for transmuting actinide wastes[27] utilized what was considered to be a more favorable direct fission-to-capture ratio for waste actinide fusion with higher energy (keV-MeV) neutrons. Extensive studies of these ideas were disappointing from several points of view, including the large waste inventory that had to be in the active system to sustain the reaction rate, the tendency to generate new and equally offensive waste products, and the inability to handle all kinds of waste products (both actinides and fission products) in the same system.

The new method indicates positive answers to all these concerns. Actinide wastes are burned or used as fuel, undergoing transmutation to stable or short-lived products with halflives on the order of 100 years or less, along with burning of fission product wastes. The system has the potential for disposing of existing wastes in a reasonable number of facilities over a period of ~100 years (both figures of 100 years are important, as comparable to a human lifespan, rather than anticipating management of such wastes by society over a longer time); in other words, reduction of both the radioactivity and the volume of the waste to low levels over the span of a human life, rather than leaving the waste to decay naturally over periods of thousands to millions of years.

A thermal neutron flux in the 10^{16} n/cm-sec range is used, enabled by the practicality of the 1.5 GeV, 250 mA class high-brightness proton accelerator discussed above. The proton beam hits a flowing liquid lead target, generating about 55 neutrons per proton. An immediate advantage is that only about 30 MeV of proton energy is deposited per neutron produced, compared to about 200 MeV of deposited energy per useful neutron in a reactor. The flowing target can also handle a larger power density that can a fixed fuel reactor system, so a larger flux of neutrons can be handled.

The system is outlined in Figure 2.

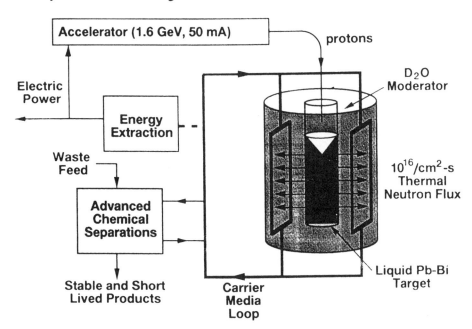

Figure 2. The accelerator-driven energy production and transmutation of waste system (ATW).

The target is surrounded by a primarily D_2O blanket to moderate the neutrons to thermal energy. Actinide material is carried continuously through the blanket as a dilute solution in a carrier material -- preferably molten salt in the energy-production configuration where the high thermal temperature of the molten salt would give high thermal-to-electric conversion efficiency. The burnup rates are very high, so the continuous flow system is very practical from this aspect as well, compared to fixed, batch-type fuel assemblies. Advanced chemistry processes on the flowing streams remove stable or short-lived fission products and return radioactive waste to the blanket again. Part of the electricity produced is fed back to power the accelerator. Efficiency calculations show that electrical power can probably be produced at presently competitive rates, especially if the molten-salt carrier is proven out, with costs calculated as done at present without factoring in any worth for the absence of the long-term waste stream.

The basic advantage of the transmutation physics using thermal neutrons is that, in a high enough neutron flux, two neutrons are captured in succession by a target actinide nucleus, and fission occurs, producing about 2.7 neutrons, so the actinide behaves as a net neutron producer or fuel. In a lower flux, the actinide nucleus captures one neutron and then goes through a decay process before finally fissioning, requiring on the average about four neutrons input for the fission release of around 2.9 neutrons, so in this case the actinide behaves as a neutron sink, or poison. This inefficiency for a low intensity thermal neutron flux is what focused attention earlier on fast flux systems, where the fission probability is more favorable although the small cross sections require a large inventory. A desired transmutation rate is governed by three factors; the neutron flux, the cross section, and the mass of the inventory. In our system with very high thermal neutron flux, enabling the efficient two-step capture-fission reactions, a smaller fuel inventory is possible. The studies indicate that the accelerator enables the following important new features in a thermal-neutron driven fission power system:

• The large number of neutrons provided by the accelerator allow operation well below criticality and instant shutdown.

• An increase over reactor systems up to 50% in the effective average number of neutrons per fission, with the enhancement adjustable to the desired application.

• The target advantages of less heat generated per neutron produced, and flowing medium, enabling higher flux production capability and chemical treatment options.

• Use of the fact that the flux production of up to 100 times that of other systems enables a reduction by the same factor in the time required to transmute fission product isotopes with low capture cross sections.

• Use of the previously unrecognized process sketched above whereby higher actinides are converted from poison to fuel in the high thermal flux.

• Reduction of the resident inventory required, compared to using keV/MeV neutrons, by a factor of about 100 for fission products because of the high thermal neutron flux from the accelerator-driven source; and by a factor of about 1000 for higher actinides from both the high flux and the high thermal cross sections.

With this system, it may be practical to realize both the Th/U and U/Pu fuel cycles for energy production, thus realizing an energy supply from a very large and available fuel supply, and without a waste stream requiring long-term management.

The accelerator required in a point-design 3000-MWt energy-production system is about 93 mA average proton current at 1.6 GeV, or 220 mA averge at 800 MeV. A room temperature 1.6 GeV, 250 mA 100% duty factor design was proposed[18] to the DOE for accelerator production of tritium (APT), another application, and favorably reviewed by the DOE Energy Advisory Review Board, stating *"The continuous-wave RF-linac approach for APT is technically sound. While an integrated accelerator system has not been built and operated at APT conditions, the accelerator feasibility and engineering development issues could be resolved with an adequate research, component and systems development, and engineering demonstration program.*"

We believe that a superconducting cw linac with a real-estate gradient of only 3 MeV/m would have lower capital cost and higher operating efficiency than the room temperature linac[28]. This clearly seems the route of the future, but a development program is needed to produce fully-engineered superconducting structures at the full range of particle velocities in the 1.6 GeV linac, to answer questions about possible Q degradation from hydrogen monolayer formation, development of control and maintenance procedures, and so on.

Intense Neutron Source for Basic and Fusion Materials Development

Interest in a facility for materials testing and development has perked up again[30-32], centered on a 35 MeV cw deuteron linac providing up to 250 mA per module, with advanced features for energy variability and shaping capability of the beam on target to tailor the flux distribution in the test cell. JAERI in Japan has proposed an Energy Selective Neutron Irradiation Test (ESNIT) facility at the 50 mA cw level. The accelerator requirements are similar to those for the applications above.

REFERENCES

1. Proceedings of an Information Meeting on Accelerator-Breeding, BNL, January 18-19, 1977, Conf-770107.
2. 1975 Particle Accelerator Conference, Wash. D.C., March 12-14, 1975, IEEE trans. NS, Vol. NS-22, No.3, June 1975.
3. R.L. Martin, "The Ion Beam Compressor for Pellet Fusion", 1975 PAC, op.cit., p. 1763.
4. A.W. Maschke, "Relativistic Heavy Ions for Fusion Applications", 1975 PAC, op.cit., p. 1825.
5. R.A. Jameson & R.S. Mills, "Factors Affecting High-Current, Bright Linac Beams", LASL Memorandum, MP-9, April 8, 1977.
6. W.P. Lysenko, "Equilibrium Phase Space Distributions and Space Charge Limits in Linacs", LA-7010-MS, LASL, October 1977.
7. F.J. Sacherer, "RMS Envelope Equations With Space Charge, CERN Internal Rpt. CERN/SI/INT DL/70-12, and same title in 1971 PAC, March 1-3, 1971, Chicago, IEEE Trans. NS, Vol. NS-18, No. 3, p. 1105.
8. M. Reiser, "Periodic Focusing of Intense Beams", Particle Accelerators, 8(3):167.
9. Tzu, Sun, "Art of War", Oxford U. Press, Oxford, UK, 1971.
10. Space Charge In Linear Accelerators Workshop, 10/31-11/2/1977, LASL LA-7265-C issued May 1978.
11. R.A. Jameson, "High-Intensity Deuteron Linear Accelerator", 1979 PAC, IEEE Trans. NS, Vol. NS-26, No. 3, p.2986.
12. R. A. Jameson, R. S. Mills, O. R. Sander, "Report on Foreign Travel - Switzerland", LASL Office Memo AT-DO-351(U)MP-9, Dec. 28, 1978; and R. A. Jameson, ""Emittance Growth in the New CERN Linac - Transverse Plane Comparison between Experimental Results and Computer Simulation", LASL Office Memo AT-DO-377(U), Jan. 15, 1979; and R. A. Jameson, "CERN Linac Tests, LASL Office Memo MP-9/AT-DO-(U), Mar. 1, 1979; and R. A. Jameson, "CERN Linac Tests" LASL Office Memo AT-DO-514(U), Apr. 26, 1979.
13. P. Lapostolle, C. Taylor, P. Tetu, L. Thorndahl, "Intensity Dependent Effects and Space-Charge Limit Investigations on CERN Linear Injector and Synchrotron", CERN Report 68:35, 1968.

14. R. A. Jameson, "Beam-Intensity Limitations in Linear Accelerators," Proc. 1981 Particle Accelerator Conf., Washington, DC, March 11-13, 1981, lEEE Trans. Nucl. Sci. 28, p. 2408, June 1981.
15. R.A. Jameson, "LINACS", a linac modeling and costing code using the rms theory as the accelerator technical model. Under construction, 1991. Written in *Mathematica*.
16. J. Struckmeier, J. Klabunde, M. Reiser, "On the Stability and Emittance Growth of Different Particle Phase-Space Distributions in a Long Magnetic Quadrupole Channel," Particle Accelerators, 15, (1984), 47.; and, T.P. Wangler, K.R. Crandall, R.S. Mills, and M. Reiser, "Relation Between Field Energy and RMS Emittance in Intense Particle Beams, 1985 PAC, IEEE Trans. NS, Vol. NS-32, No. 5, October 1985.
17. "Accelerator Production of Tritium", Presentation to the Energy Research Advisory Board, October 25, 1989, Los Alamos - Brookhaven National Laboratories.
18. O.A. Nezhevenko, Institute of Nuclear Physics, Novosibirsk, private communication.
19. e.g.: "The 2nd International Symposium on Advanced Nuclear Energy Research - Evolution by Accelerators, January 24-26, 1990, Mito, Japan, Proceedings by Japan Atomic Energy Research Institute, Tokai, Ibaraki, Japan.
20. International Symposium on Heavy Ion Inertial Fusion, Dec. 3-6, 1990, Program & Abstracts, LBL-29779. Proc. to be published by LBL.
22. "Heavy-Ion Fusion Accelerator Research 1989", LBL-28003, UC-411, June 1990.
23. "High Energy Density in Matter Produced by Heavy Ion Beams, GSI-90-15, October 1990.
24. Fusion Policy Advisory Committee, 1990.
25. I. Hofmann, "Recent Developments in Inertial Fusion Based on RF Accelerators", 1990 Linac Conf., Sept. 10-14, 1990 Proc. LA-12004-C, Los Alamos National Laboratory, p. 289.
26. "A Los Alamos Concept for Accelerator Transmutation of Waste and Energy Production (ATW)", December 10-12, 1990 External Review, LA-UR-90-4432, Los Alamos National Laboratory.
27. C.D. Bowman, E.D. Arthur, et. al., "Nuclear Energy Generation and Waste Transmutation Using An Accelerator-Driven Intense Thermal Neutron Source", Los Alamos National Laboratory, being submitted for publication.
28. International Conference on Nuclear Waste Transmutation, July 22-24, 1980, The University of Texas at Austin, College of Engineering, Center for Energy Studies, Proc. publ. March 1, 1981.
29. G.P. Lawrence, "Accelerator Issues and Development for Defense-Waste Transmutation and Energy Production", presented to AT-Division Advisory Board, May 16-17, 1991, Los Alamos National Laboratory.
30. "High Energy Neutron Source for Material Research & Development", Tokyo, Japan, Jan. 12-13, 1989, Proc. by Japan Atomic Energy Research Institute, Tokai, Ibaraki, Japan, March 1989.
31. Nuclear Science and Engineering, Vol. 106, Nos. 2 & 3, October and November 1990, American Nuclear Society; special issues devoted to high-intensity fast neutron sources and calibrated neutron fields for fusion technology and fusion materials research.
32. R. A. Jameson, "Energy Variable Deuteron Linac For Materials Research Neutron Source", Proc. 1990 2nd European Particle Accelerator Conference, Nice, France, 12-16 June 1990, p.1837.

NON-LIOUVILLEAN METHOD APPLIED TO HEAVY ION FUSION

I. Hofmann

GSI Darmstadt, 6100 Darmstadt, Germany

ABSTRACT

Heavy ions are a challenging option for ICF driver accelerators. We focus on the rf linac / storage ring approach and an advanced scheme for indirectly driven targets. The main emphasis is on an increase of phase space density by means of a non-Liouvillean scheme using photoionization of single charged heavy ions. The required large cross sections are known to exist for Ba^+ stripped to Ba^{++}. In contrast with the non-Liouvillean injection methods proposed by previous authors, we discuss here photoionization for *extraction* from the storage ring into a compression ring, where the beam stays only a few revolutions. This promises a significant improvement of accelerator performance. Design examples for a full driver scenario as well as an experimental facility leading to an ignition experiment are presented. Experimental results pertaining to the longitudinal microwave instability as one of the key issues are reported and related to the design of an advanced driver accelerator scheme.

INTRODUCTION

One of the major conclusions of HIBALL [1] has been that a uniformity of target illumination of about 1% (necessary to make a directly driven target work [2]) cannot be realized in a reactor, where for design reasons only beams hitting the target under relatively small angles from the equatorial plane are considered. A separation of the problem of target implosion symmetry from the symmetry of beam illumination can be achieved with indirectly driven pellets in a hohlraum cavity. The conversion of beam energy into soft x-radiation requires an enhanced specific power, which has been estimated to be 10^{16} W/g [3, 4]. This specific power, which can be achieved by a smaller focal spot area, is an order of magnitude higher than in HIBALL, which has assumed phase space densities already close to theoretical limits.

One way of dealing with this challenging problem is the introduction of a non-Liouvillean scheme, which allows a real increase of the phase space density. After successful demonstration in the late 60's of H^- stripping injection at Novosibirsk, and later at Argonne, such a scheme was proposed in the early U.S. program on heavy ion fusion to accumulate, without increase of phase space, the beam from a low-current linac. One of the schemes proposed in 1976 was photo-dissociation of iodine hydrogen by flashlamp or laser light (see Ref. [5]). After it was realized that high-current heavy ion sources could be built, non-Liouvillean injection was decided to be unnecessary. Recently the idea was revitalized by Rubbia who suggested photoionization of Bi^+ [6] at injection from the linac into the storage ring in order to replace the complicated horizontal and vertical multiturn stacking used in HIBALL.

A significant improvement of the final phase space density can be achieved by shifting the non-Liouvillean scheme from the injector side to the extraction from the storage ring into the bunch compression ring ("non-Liouvillean bunch compression", see Ref. [7]). This requires an ion with large resonant autoionization cross section, like Ba^+ stripped to Ba^{++} so that ionization can occur over a short common straight section of the two rings [7].

Another important issue related to the high currents is final bunching in the presence of space charge. In order to avoid unconventionally large buncher voltages one can work with a sufficiently large (typically 10^3) number of beamlets. This leads to the concept of "bundled" beams, where a bundle of several 10's of beamlets goes through a common focusing structure. The idea of matrix arrays has first been proposed by Maschke [8] in his MEQALAC scheme for electrostatic quadrupole arrays. Basic parameters are summarized in Table I for an ion with the mass of Bi, which is also used in the rest of the paper as a reference ion for the accelerator parameters (assuming the existence of an ion with a mass near 200 and a cross section comparable to that of Ba^+).

Table I. Parameters for Indirectly Driven Fusion.

Ion	$A \approx 209$ (stripped + to ++)
Kinetic energy	10 GeV
Total energy	5 MJ ($N = 3.1 * 10^{15}$)
Final pulse duration	20 ns (full width)
Final momentum spread	$\Delta p/p = 3.0 * 10^{-4}$
Emittance at target	6π mm mrad
Spot diameter	2.7 mm
Ion range	0.3 g/cm^2
Specific Power	10^{16} W/g (maximum)

NON-LIOUVILLEAN SCHEME

The general layout of a scheme with the above features is shown in Fig. 1. The current multiplication from linac to storage ring is performed by 4-fold horizontal multiturn injection in each transfer ring. After filling of all 4 rings (140 μsecs) their content is transferred into one storage ring by 4-fold vertical multiturn injection. This is repeated 20 times until all storage rings are filled (3 msecs). The storage rings are separated rings arranged vertically in stacks of 4 rings each.

For an emittance of 5π mm mrad we obtain a Laslett tune shift $\Delta Q = 0.37$ for the coasting beam, which should be tolerable for the assumed maximum storage time during filling of all the rings within 5...10 milliseconds. After filling is completed the beam is adiabatically bunched on harmonic 240 leading to 20 nsec long bunches.

The technically feasible maximum rf voltage in the rings is a major factor in the design as one cannot have large voltage and low frequency at the same time. Since the voltage is more determined by space charge than by momentum spread, it has to be adjusted to the increasing current during the non-Liouvillean stacking. In our case we require a maximum of 16 MV/turn at 28 MHz.

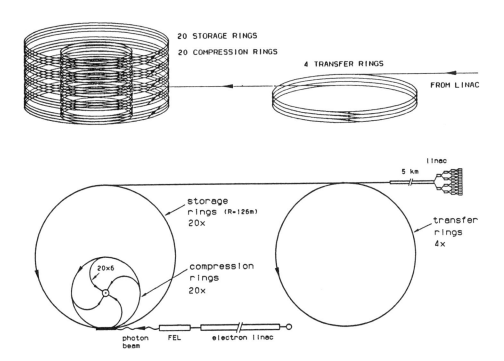

Fig. 1. Schematic Layout of an Advanced Driver Scheme.

In the following we discuss some general features of the proposed non-Liouvillean compression scheme. The photon beam from a free electron laser is turned on for an interaction time Δt_i corresponding to the bunch length, during which the charge state is changed from 1+ to 2+ in a single transit. The simplest scheme (Fig. 2) is using two strong dipole magnets to separate the orbits.

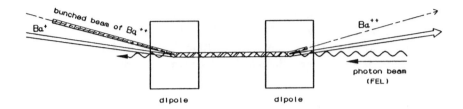

Fig. 2. Scheme of photoionization non-Liouvillean procedure

Due to the doubled magnetic rigidity and by using superconducting magnets it is possible to introduce an angle of the order of 100 mrad between the two beams and thus guide the doubly charged ion into the adjacent compression ring.
Due to the change of charge state the phase space trajectory of a bunch entering the compression ring is merged with that of an already orbiting bunch. Hence the intensity is increased without change of the phase space volume (see Fig. 3).

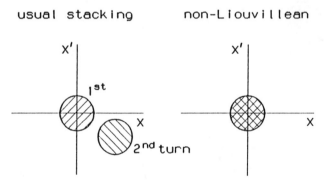

Fig. 3. Schematic view of phase space

This procedure is repeated ten times over ten revolutions, hence ten bunches of the storage ring are thus merged into a single bunch in the compression ring. By appropriate timing of the laser pulse one thus obtains 24 stacked bunches in each compression ring. The required photon flux for conversion in a single pass through the interaction region depends on the detailed design of the interaction insertion and on the photoionization cross section.

We first consider the photoionization process $Bi^{1+} \to Bi^{2+}$ (see Ref.[6]) and assume a cross section σ_{ph}, a density of photons n_{ph}, and a conversion path length Δl. With $\Delta N = N n_{ph} \sigma_{ph} \Delta l$ we find for 90% conversion that

$$n_{ph} \sigma_{ph} \Delta l \approx 2.3 \tag{1}$$

We thus assume an interaction length of 10 m and obtain

$$n_{ph} \approx 2.3 * 10^{-3}/\sigma_{ph} \quad (cm^{-3}) \tag{2}$$

The β - functions in the drift section of our lattice are assumed to be typically 10 m, hence for an emittance of 6 π mm mrad we obtain a typical beam cross section of $F \approx 2\ cm^2$. This enables us to calculate the power of the photon beam for photons with energy E (eV) given by

$$P - 2.3 \frac{FceE}{\Delta l \sigma_{ph}} \quad (W) \tag{3}$$

which results for our example as:

$$P \approx 2. * 10^{-11} E/\sigma_{ph} \quad (W) \tag{4}$$

In the moving frame we require a Doppler shifted photon energy of 20 eV for $Bi^{1+} \to Bi^{2+}$ and in the laboratory frame a Doppler shifted energy of 14.5 eV. If we assume a cross section of $3 * 10^{-17}\ cm^2$ this results in a laser power of 10 Megawatts to be delivered during the pulse time of 20 ns.

This laser power might be higher than technically feasible and we have therefore looked at other ions with possibly more favourable cross sections. Measured cross sections exist for Ba^+ ions[9]. For energies above the ionization threshold, namely at about 21.2 eV, these cross sections are as large as $2.8*10^{-15}\ cm^2$. They are sharply peaked and due to resonant autoionization effects. Simple estimates show that the energy and angular resolution of our beams is roughly compatible with the resonance width, which is about 0.025 eV. For this case there could be a reduction of the required laser beam power by almost two orders of magnitude, hence 200 kW would be sufficient. In the literature we have not found information about cross sections for further ionization due to the repeated traversal of the doubly charged ions through the laser beam. These should be at least 2-3 orders of magnitude lower, which is realistic in view of the large resonant values for the transition 1+ to 2+.

We estimate that the required laser power can be delivered by a FEL driven by an electron linac. The design of such a device requires a separate study.

An important consequence of the relatively large number of beamlets (here 480) is the relaxation of space charge issues. They can be guided to the reactor not by individual beam lines, but in bundles with matrix focusing magnets for each bundle [10].

No extra rf bunching is required outside the compressor rings. Provided the transport lines to the reactor are sufficiently short (a few hundred meters) it is possible to counteract the space charge debunching by a moderate additional voltage applied in the compression rings prior to ejection. It is essential for this scheme to have delay lines, which make all bunches arrive at the target simultaneously. To keep these delay lines short one can place the stacked 24 bunches of each compressor ring not at equal distances, but into subsequent buckets. One thus obtains a train of length 24/120=1/5 th of the circumference and hence an equally long maximum delay.

Parameters for the scheme shown in Fig. 1 are summarized in Table II.

Table II. Parameters of Design Example.

linac: length	5 km
maximum current	190 mA
emittance (mm mrad)	0.5π
momentum spread	$3\ 10^{-4}$
transfer rings (4): radius	126 m
multiturn injection (h)	4 x
emittance (mm mrad) (h/v)	$5\pi\ /\ 0.5\pi$
space charge tune shift (h/v)	0.14 / 0.42
storage rings (20): radius	126 m
multiturn injection (v)	4 x
emittance (mm mrad)	$5\pi\ /\ 5\pi$
space charge tune shift	0.37 (coasting)
bunching rf	10 MV/turn
bunch length	20 nsec
rf frequency	28 MHz
compression rings (20): radius	63 m
non-Liouvillean stacking	10 x
emittance (mm mrad)	6π
tune with max. space charge	$Q_0/3$
bunching rf	16 MV/turn
final transport: distance	100 m
emittance (mm mrad)	6π
momentum spread	$3\ 10^{-4}$

I. Hofmann   155

The funneled linac follows closely the HIBALL design, but with less demands in phase space density. It is thus appropriate to assume a driver efficiency of 26 % as in HIBALL, which is the main merit of heavy ion accelerators and lends itself to economical energy production.

EXPERIMENTAL FACILITY

In this section we discuss a set of parameters of an experimental accelerator device which is based on the same scheme as the above full driver scenario and lies within the same fundamental beam dynamics limitations. It is assumed that such a facility is first built on a smaller scale, with a shorter linac leading to 1 GeV ions and only one out of 5 stacks of storage and compression rings (see Fig. 4).

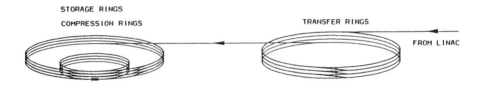

Fig. 4. First stage of an experimental facility.

In two following stages more storage rings can be added and the length of the linac increased to an energy of 3 GeV.

In the upgraded third stage the energy leading to ignition experiments should be reached. We have found that a kinetic energy of 3 GeV is appropriate, if 5 stacks of storage and compression rings are used as in a driver. Keeping the same ring radius and working with superconducting dipoles, the higher energy requires an increase of the dipole field in all rings from 2.9 T to 5 T. Parameters of these two stages for an ion with A = 200 are summarized in Table III.

RELATED BEAM DYNAMICS EXPERIMENTS

A major issue in the design of a HIF driver is the stability of the beams in the storage rings until all rings are filled. This is the longest time scale involved and it is crucial to operate on the safe side with respect to the longitudinal microwave instability. Recent experiments with cooled beams in the ESR have basically settled this issue. With combined electron cooling and rf stacking in the ESR [11]

Table III. Parameters of an experimental facility.

	First Stage	Second Stage	Third Stage
total energy	20 kJ	100 kJ	1 MJ
kinetic energy	1 GeV	1 GeV	3 GeV
linac:			
length	0.5 km	0.5 km	1.5 km
maximum current	40 mA	40 mA	130 mA
emittance (mm mrad)	6 π	6 π	0.5 π
momentum spread	3 10^{-3}	3 10^{-3}	2 10^{-3}
transfer rings :	3	3	4
radius	68.6 m	68.6 m	68.6 m
stacking	3x	3x	4x
storage rings :	4	20	20
radius	68.6 m	68.6 m	68.6 m
stacking	3x	3x	4x
compression rings :	4	20	20
radius	34.3 m	34.3 m	34.3 m
non-Liouv. stacking	10x	10x	10x
final on target:			
emittance (mm mrad)	30 π	30 π	20 π
momentum spread	3 10^{-3}	3 10^{-3}	2 10^{-3}
number of beamlets	480	480	480
bunch length	20 nsec	20 nsec	20 nsec

we have obtained phase space densities where beam dynamics is influenced by collective effects similar as in HIF storage rings.

The implications of these results for heavy ion fusion driver design are important as they show that the usual asumption of the conventional Keil-Schnell threshold is by far too pessimistic.

Details of the diagnostics set-up are described in Ref. [12, 13]. The Schottky spectrum for an electron cooled beam of Ar^{18+} at 250 MeV/u is shown in Fig. 5 for an initial $(\Delta p/p)_{fwhm} = 10^{-3}$. The origin of the Schottky spectrum is the noise from the statistical distribution of particles. This gives rise to current fluctuations, which induce a voltage on a pick-up. The cooling equilibrium depends on the electron current. For $I_e = 0.1$A the spectrum shows the momentum distribution, whereas for $I_e = 1$A the double peak spectrum reflects anomalous behaviour due to collective effects. In the latter case the momentum spread was reduced to 2.6 10^{-5}. The ion current has been 1 mA (electric), which was obtained by "cooling stacking".

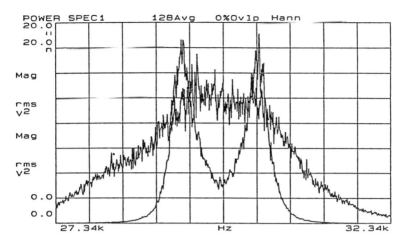

Fig. 5. Schottky Spectrum for $I_e=0.1$ A and 1 A.

It is interesting to express the result for cooled bunches in terms of the factor, by which the actual current exceeds the Keil-Schnell circle threshold:

$$\frac{I}{I_{KS}} = \frac{4eqIZ_\| \ln 2}{\pi|\eta|^2\beta^2\gamma Amc^2(\frac{\Delta p}{p})^2_{fwhm}} \qquad (5)$$

with $Z_\|$ the coupling impedance, $\eta = 1/\gamma^2 - 1/\gamma_t^2$ and $(\frac{\Delta p}{p})_{fwhm}$ the fwhm momentum spread. This dimensionless factor can be regarded as "figure of merit" of longitudinal cooling [14].

In the HIBALL study it was assumed that due to stabilizing tails the factor I/I_{KS} can be as large as 50, without loosing stability with respect to the longitudinal microwave mode. In our design for an advanced driver scenario it has been possible (owing to the non-Liouvillean technique) to keep this factor below 10. The present measurements, supplemented by results from other rings at different energies give experimental confirmation that for coasting beams such a factor can indeed be achieved (Fig. 6).

Of particular interest in this context are results from the TSR Heidelberg with C^{6+} at lower energy, which have shown that the excess factor can be as large as 6-10, depending on the intensity [15]. The comparison between their 2.5 mA and 15 mA case supports the asumption that intrabeam scattering prevents a large excess factor if the intensity is too low. For HIF, on the other hand, intrabeam scattering is unimportant as we deal with single or double charged ions (noting that the intrabeam scattering rate scales with Z^4).

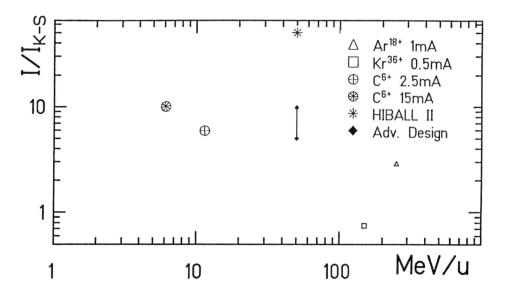

Fig. 6. Comparison of Measured I/I_{KS} with Heavy Ion Fusion.

SUMMARY

An advanced driver for indirectly driven targets has been examined on the basis of non-Liouvillean stacking. It helps to achieve maximum phase space densities at the very last stage of the scheme and leads to comfortably reduced phase space densities in the transfer rings and storage rings. The beam dynamics issues and atomic physics issues involved in photoionization stacking as well as the design of an appropriate FEL with 10^5 to 10^6 W of VUV light require careful studies and experimental verification. Parameters of an experimental facility for accelerator demonstration as well as significant target physics have been discussed. A first stage at 1/250 th of the total energy output of a full driver requires a linac of 1/10 th of the driver linac and 1/5 th of the number of storage rings. Upgrading to 1/3 rd of the driver linac and the full number of storage rings gives an output of 1 MJ, which could lead to ignition experiments.

Recent measurements on the threshold of the longitudinal microwave instability have been shown to give an experimental basis for the design of the storage rings involved in the advanced driver. The most important result is that the conventional Keil-Schnell-factor can be exceeded by a factor 5-10. Further experiments related to space charge in bunched beams are underway.

References

[1] B.Badger et al., Karlsruhe Report KfK-3480 (1984)

[2] G. Buchwaldt et al., Laser and Particle Beams 1, 335 (1983)

[3] M. Murakami, J. Meyer-ter-Vehn and R. Ramis, Journ. of X-Ray Science and Techn. 2, 127 (1990)

[4] J. Meyer-ter-Vehn, "Analysis of Heavy Ion Fusion Targets", Proc. Int. Symp. on Heavy Ion Fusion, Monterey, Dec. 3-6, 1990

[5] R.L. Martin, "History of Non-Liouvillean Injection and its Connection to the Initiation of the U.S. Program in Heavy Ion Fusion", these proceedings;
R. Arnold, R. Burke, Y. Cho, R. Cutler, S. Fenster and R.L. Martin, IEEE Trans. Nucl. Sci. NS-24, 1428 (1977)

[6] C. Rubbia, Nucl. Instr. and Meth. A278, 253 (1989)

[7] I. Hofmann, Laser and Particle Beams 8, 527 (1990)

[8] A.W. Maschke, Brookhaven National Laboratory Report, BNL 51029, 1979

[9] Lyon et al., J. Phys. B, At. Mol. Phys. 19, 4137 (1986)

[10] C. Rubbia, "On Heavy Ions Accelerators for Inertial Confinement Fusion", Proc. IAEA Technical Meeting on Drivers for Inertial Confinement Fusion, April 15-19, 1991, Osaka

[11] B. Franzke et al., Proc. European Part. Accel. Conf., Nice, June 12-14, 1990, p.46

[12] U. Schaaf, "Schottky Diagnose und BTF-Messungen an gekühlten Strahlen im ESR", Dissertation, Frankfurt University (1991)

[13] K. Beckert, S. Baumann-Cocher, B. Franzke and U. Schaaf, Proc. European Part. Accel. Conf., Nice, June 12-14, 1990, p.777

[14] I. Hofmann, K. Beckert, S. Baumann-Cocher and U. Schaaf, Proc. European Part. Accel. Conf., Nice, June 12-14, 1990, p.229

[15] TSR group, "First experiments with the Heidelberg Test Storage Ring TSR", MPI H - V18 (1989)

MBE-4 EXPERIMENTS WITH BRIGHT CESIUM+ BEAMS[†][*]

T.J. Fessenden

Lawrence Berkeley Laboratory, Berkeley, California

SUMMARY

Since 1985 the Heavy Ion Fusion Accelerator Research program at the Lawrence Berkeley Laboratory has been studying current amplification and emittance variations in MBE-4, a four-cesium-beam ion induction linac. This experiment models much of the accelerator physics of the electrostatically focused section of a fusion driver. Four space-charge dominated Cs+ beams, initially about one meter in length at currents of 5-10 mA, are focused by electrostatic quadrupoles and accelerated in parallel from approximately 200 keV up to one MeV by 24 accelerating gaps. Final currents of 20-40 mA per beam are typical. Recent experiments with extremely low emittance beams ($\epsilon_n = 0.03$ π mm-mRad) have investigated variations of transverse and longitudinal normalized emittance for drifting and accelerating beams. Experiments show that very cold ($\sigma_0 = 72°, \sigma = 6°$), off-axis or poorly matched beams increase transverse emittance when drifted or accelerated through these MBE-4 apparatus. Only by carefully centering and matching the beams can acceleration at constant normalized emittance be achieved. Warmer beams with less tune depression exhibit little to no emittance growth and show smaller emittance fluctuations when off axis or mis-matched.

INTRODUCTION

The Heavy Ion Fusion Accelerator Research Program (HIFAR) at LBL is assessing the multiple-beam induction linac as a inertial fusion driver. In this concept multiple parallel beams of heavy ions are continually amplified in current and in voltage as they are accelerated to the parameters required to ignite an inertial fusion target ($\approx$10 GeV, 500 TW, 10 ns). Control of the lengths of the beam bunches during the acceleration process is one of the key beam dynamics issues of this approach to a fusion driver. The accelerating waveforms must be carefully shaped to shorten the bunch length and to control longitudinal space charge forces while accelerating the beams. Small acceleration errors, particularly troublesome at low beam energies, can lead to current spikes and beam spill and/or to unacceptable increases in beam emittance. A necessary consequence of current-amplifying acceleration is that the focusing system must transport beams whose speed at a focusing element increases by as much as 20% over the time of the pulse.

MBE-4 is a multiple-beam current-amplifying ion induction accelerator. It was built to develop an experimental understanding of this new type of accelerator. Longitudinal beam dynamics in this experiment are similar to those expected in the electrostatic-focused region of a heavy ion driver. Experiments began with the completion of the injector in 1985 and continued for approximately six years through April 1991. Of particular interest were possibilities of interactions between the multiple

---

[†]Much of this work was presented at the 1991 Particle Accelerator Conference in San Francisco as paper GGC8

[*]Work supported by the Office of Energy Research, Office of Basic Energy Sciences, U.S. Department of Energy under Contract DE-AC03-76SF00098.

Fig 1 The MBE-4 multiple-beam ion accelerator (CBB 912-1174)

beams, longitudinal beam control, and the preservation of longitudinal and transverse normalized emittance during acceleration. The main thrust of the first three years of experimentation was longitudinal beam dynamics and control. The last three mostly studied transverse beam control and emittance growth during acceleration. MBE-4 has been very useful in understanding the physics of current-amplifying accelerators and in providing confidence that much larger ion induction linacs can be expected to operate satisfactory. This paper will summarize our findings from MBE-4.

DESCRIPTION OF MBE-4

MBE-4 is an induction linac that accelerates four Cs^+ beams from 200 keV at the injector to as much as 875 keV(head); 950 keV(tail) after 24 accelerator gaps with current amplifications as large as nine. Transverse focusing is supplied by four-beam arrays of electrostatic quadrupoles. The initial bunch length is 1.3 m with duration of 2.2 µs. Fabrication of the apparatus was completed in September 1987. A picture of MBE-4 is presented in Fig. 1 and a schematic of the accelerator is given in Fig. 2. The accelerator is 30 periods long. Pumping and diagnostic access are provided at the ends and at each 5th focusing period. Our principal diagnostics are Faraday cups and two-slit scanners which reveal the beam size and emittance for current measurements at each of the diagnostic stations. An electrostatic energy analyzer at the end of the experiment is used to measure the final beam energy and obtain estimates of the longitudinal beam emittance. We also used a small electrostatic energy analyzer that could be inserted at the diagnostic stations to characterize the beam at the input and along the accelerator. During operation the apparatus is pulsed every five seconds. Each pulse is highly repeatable. Every waveform repeats to better than 1 %.

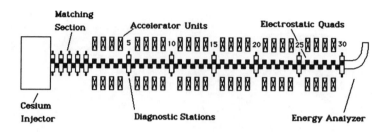

Fig. 2 Schematic of MBE-4.

LONGITUDINAL STUDIES AND FINDINGS

Current amplification and longitudinal bunch length control through the accelerator require the use of carefully shaped accelerating voltage waveforms. The method of finding waveforms for accelerating beams in ion induction linacs, was developed by C. Kim[1]. In this current self-replicating scheme, the functional form of the current versus time at a fixed location is preserved through the accelerator and the magnitude increases as the bunch shortens in time. Solutions for the current and the accelerating waveforms at every accelerating gap can be constructed. The charge distribution along the length of the beam bunch generates a longitudinal electric field E_z that will lengthen the bunch if not compensated by the accelerating voltages. The procedure for generating these wave forms including the effects of longitudinal space charge and the finite width of the accelerating gaps were incorporated into a code called SLID which runs on a small computer.

The SLID procedure has proved to be extremely valuable for engineering pulsers for MBE-4 and for understanding and interpreting the results of the experiments. An improved version of the procedure called SLIDE[2] which permits particle overtaking was recently developed by Henestroza. We have found that the SLID and SLIDE calculations of the current and energy waveforms agree very well with experimental measurements[3,4].

To develop the pulsers for controlling and accelerating beams in MBE-4, ideal waveforms from the SLID procedure were first supplied to engineering. As the pulsers were fabricated, the outputs of the pulsers were measured and used as input to the SLID code to generate downstream waveform requests that tended to compensate for unavoidable upstream synthesis errors. Our method of synthesizing waveforms was to add the outputs of several pulsers so as to generate the waveform asked for by the SLID procedure as well as possible. An example of this synthesis is presented in figure 3. Here the outputs of four pulsers are added at gap 11 to generate the waveform requested by the SLID procedure. As can be seen, small errors are inevitably generated as each pulser is energized. Waveforms to provide control of the bunch ends are not present at each accelerating gap but rather are provided at every five to seven gaps.

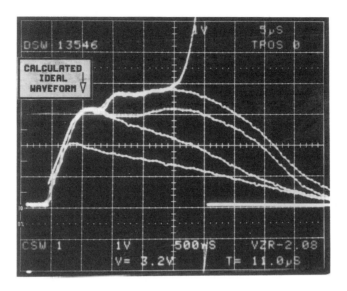

Fig. 3. Synthesis of MBE-4 accelerator wave forms. The outputs of four pulsers are added to produce an accelerating voltage close to that requested by the SLID acceleration procedure (XBB 872-1623)

Amplifying current waveforms obtained in MBE-4 are shown in Fig. 4. Here the current of each of the four beams increases from 10 mA at 0.2 MV at the accelerator input to over 90 mA at end of the accelerator. At exit the beam energy increases approximately linearly from 650 kV at the head to 750 kV at the tail. The effect of the acceleration errors can be seen in these waveforms.

TRANSVERSE STUDIES AND FINDINGS

Since approximately August of 1988, the preservation of normalized emittance with acceleration has been the principal issue studied with MBE-4. We had observed during the longitudinal experiments that the transverse emittance remained approximately constant or increased with drift or acceleration implying that the normalized emittance was increasing. Moreover, although each emittance measurement could be repeated to within about 5% if done immediately, measurements over long periods varied by factors of 2 to 3.

Figure 5 shows data collected from accelerated beams at an initial current of 10 mA per beam over a period of several months. Very similar emittance variations obtained from 2-D PIC simulations of strongly tune-depressed beams drifting in electric-quadrupole-focused transport systems had been reported previously by Celata[5]. These simulation studies were continued and extended to accelerating beams by K. Hahn. We define normalized emittance as $\varepsilon_n = 4 \beta\gamma \varepsilon_{rms}$ where ε_{rms} is given by

$$\varepsilon_{rms} = [\langle x^2 \rangle \langle x'^2 \rangle - \langle xx' \rangle^2]^{1/2} \tag{1}$$

164 MBE-4 Experiments with Bright Cesium⁺ Beams

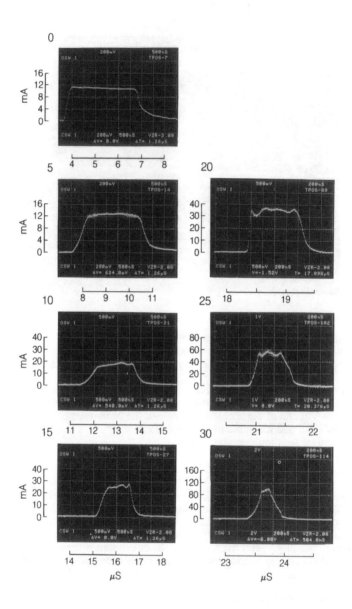

Fig 4. Current amplification from 10 to 90 mA/beam in MBE-4. Each oscillogram shows several pulses (XBB 887-7123)

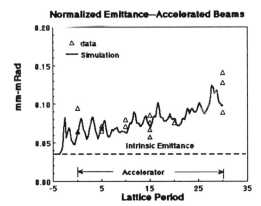

Fig. 5 Measurements and simulation of normalized emittance along MBE-4. The intrinsic emittance is that determined by the source temperature (≈ 1000°C) and beam radius.

A simulation in which the measured beam in phase space at the output of the injector was used to load the code is also presented in Fig. 5. The intrinsic beam emittance as determined by the source radius r_s and the rms thermal velocity $v_\emptyset$ at a temperature of 0.1 eV was used at the start of the simulation. The initial emittance growth is due to a space-charge-driven re-arrangement of particles in the beam[6]. Diagnostic access and, consequently, emittance measurements are limited to every 5th lattice period. Because of the sparse sampling and the rapid fluctuations in emittance, the measured emittance can both increase and decrease along the accelerator.

To demonstrate that the rms emittance was rapidly varying along MBE-4, measurements of beam offset and emittance versus the strength of the focusing lattice (s_0) at fixed position were performed. This technique (which was suggested by D. Keefe) showed that the beam was oscillating back and forth in the channel with an amplitude of 4 to 5 millimeters or nearly 20% of the channel aperture as shown in Fig. 6. This experiment also confirmed that the rms beam emittance was strongly modulated at what corresponds to 2.3 lattice periods in excellent qualitative agreement with the simulations.

Figure 7 shows the x and y normalized emittance versus z obtained from simulations of strongly tune-depressed beams in MBE-4. Except for the case of no offset, the beams were started 3 mm off axis, slightly more than 10% of the channel aperture, with equal x and y deflections. For the lower curves the simulations were started at the intrinsic beam emittance. The dots on the y emittance plot signify the emittance that might be measured at the points of diagnostic access. A simulation at twice the intrinsic emittance which is close to the final emittance of the first off-axis case is also given. These simulations suggest that strongly tune-depressed beams conserve normalized emittance if held to the system axis. Off-axis low-emittance beams exhibit large modulations in rms emittance and a net growth in passing through the transport system. At increased initial emittance, the modulations are less severe and the growth less significant. This behavior was experimentally checked[7] by approximately doubling the beam emittance at injection with a biased grid pair and measuring emittance growth through MBE-4. Reduced emittance growth was recorded as predicted by the simulation.

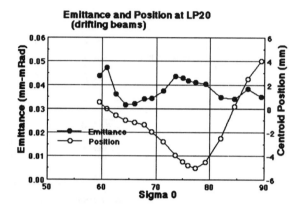

Fig. 6. Variations of beam position and emittance versus focus strength at lattice point 20. Varying focus strength brings the beam fluctuations to the point of diagnostic access.

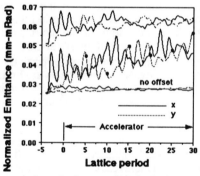

Fig. 7 Beam emittance variations in the x and y planes for offset beams drifted through MBE-4 obtained from code simulations.

Recent MBE-4 experiments have concentrated on carefully matching and centering the beams in the channel to demonstrate current amplification at constant normalized emittance. A second steering array was placed at the entrance to the accelerator that, with the first array at the injector output, centered and aligned the beam on axis. In our initial attempt[8], the beam was centered but not adequately matched. As a consequence, emittance growth was observed. Only after carefully matching, centering, and aligning the beams could these very cold beams be accelerated at constant normalized emittance. Figure 8 shows measurements of normalized beam emittance versus position and a comparison with simulation. Much more information on these experiments was presented at the 1991 Particle Accelerator Conference by T. Garvey[9] et al.

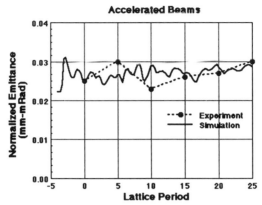

Fig. 8 Measurements and simulation of normalized emittance during acceleration through MBE-4. For these experiments the beams were very carefully matched, centered and aligned in the focusing channel.

Current amplification by drift-compression was studied in MBE-4 by using the first four accelerators to impart a voltage ramp or tilt to the beams. The beam emittance was studied as the beams drifted through the remainder of the apparatus. Figure 9 shows the current versus time at lattice point 20 for increasing initial tilts. The lowest case was obtained with no applied tilt or acceleration and shows the effects of longitudinal beam spreading due to uncorrected axial space charge forces.

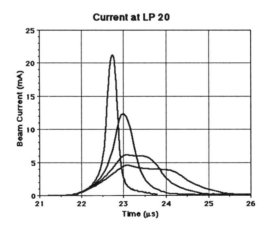

Fig. 9. Measurements of beam current at lattice point 20 for various beam tilts applied at the first four accelerating gaps. The lowest curve was obtained for no applied acceleration.

The growth in longitudinal emittance for the case of maximum current amplification is shown in Fig. 10. Both theory and experiment[10] indicate that the beam radius increases as the square root of the current. As the beam size increases, the nonlinear effects of image charges and of the contributions of higher order multipoles of the focusing fields causes the emittance to grow. For these experiments a beam loss of approximately 10 % was observed near the point of maximum emittance growth.

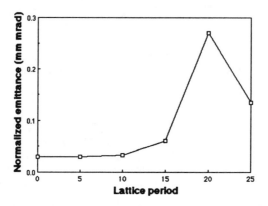

Fig. 10 Measurements of beam emittance along MBE-4 for the case of maximum current growth above.

DISCUSSION

The MBE-4 experiments were performed to obtain practical experience with a current amplifying multiple-beam ion induction linac. One of the initial concerns was possible interactions between the beams as they were accelerated or transported through the accelerator. With the exception of a weak interaction in the injectors where the beams are at low energy and "see" each other for some distance, no such effect was observed. At low energy the beams interact primarily through their radial space-charge electric fields. These are easily controlled by radial shorting planes along the accelerator which are present for other reasons. At high beam energy, magnetic interactions play a stronger role and some periodic beam coupling phenomena could conceivably arise. The major multiple-beam issue encountered in these experiments was the diagnostic complexity associated with working with many beams.

Current amplification in an induction linac is possible and the beams can be controlled longitudinally with the accelerating waveforms. However, longitudinal control in MBE-4 was more difficult than we anticipated. The accelerator pulsers used for MBE-4 are based on conventional thyratron-switched pulse forming lines. This type of pulser does not have adequate waveform flexibility to easily control the beams longitudinally. Future ion induction linacs will require more agile correction or compensating pulsers every few accelerating gaps--particularly at the low energy end.

Beam space charge greatly reduces the impact of longitudinal acceleration errors as can be seen from Eq. 3. At very low beam intensities the response to an error becomes large with time or drift length. The acceleration errors can be repaired or compensated downstream if corrections are made before the forward and backward space charge waves on the beams can split. Although these corrections contribute to the longitudinal temperature of the beams, the scaling formulas suggest that these will not unduly compromise the final focusing of the beam onto the target.

Transversely, very cold space-charge dominated ion beams can be accelerated at constant normalized emittance at least over the length of MBE 4 if great care is taken to match, center, and align these beams in the transport channel. However, for long linacs or drivers, extreme care may not be required. These experiments and simulations suggest that for sufficiently long system, the beams will center and match themselves with an accompanying increase in emittance. This emittance growth may be affordable. Further study is required (experiments, theory, simulation) to determine if this so.

ACKNOWLEDGEMENTS

The work presented here is a summary of the efforts of much of the LBL HIFAR group since 1985. Notable among these are A. Warwick, and C. Kim who made major contributions to the design of MBE-4 and to the longitudinal studies; H. Meuth, S. Eylon, and T. Garvey who concentrated more on the transverse studies; K. Hahn, E. Henestroza, L. Smith, A. Faltens and more recently W. Fawley who provided the theoretical guidance and simulation backup; and last but far from least was D. Keefe who contributed in every conceivable way to the MBE-4 experimental program.

REFERENCES

1. C.H. Kim and L. Smith, "A Design Procedure for Acceleration and Bunching in an Ion Induction Linac" Part. Accel. 85, pp 101-113.
2. E. Henestroza, "Study of the Longitudinal Ion Dynamics Extended Version; SLIDE"HIFAR Year End Report 1989
3. T.J. Fessenden, D. Keefe, C. Kim, H. Meuth, and A. Warwick, "The LBL Multiple Beam Experiments" IEEE PAC Cat. No. 87CH2387-9, Vol. 2, pp. 954-956
4 A.I. Warwick, D.E. Gough, D. Keefe, and H. Meuth,"Acceleration, Current Amplification and Emittance in MBE-4, an Experimental Beam Induction Linear Accelerator for Heavy Ions," Proc. 1988 Linear Accelerator Conf., p. 51, Oct. 88.
5. C.M. Celata, "The Effect of Nonlinear Forces on Coherently Oscillating Space-Charge-Dominated Beams" IEEE PAC Cat. No. 87CH2387-9, Vol. 2, pp. 996-1000.
6 T.P. Wangler, K.R. Crandall, R.S. Mills, and M. Reiser, "Relation between Field Energy and RMS Emittance in Intense Particle Beams," IEEE Trans. on Nucl. Science, NS-32, (1985)
7. S. Eylon, E.R. Colby, T.J. Fessenden, T. Garvey, K. Hahn and E. Henestroza, "Emittance Variations of Very Cold Ion Beams During Transport Through MBE-4," to be published in Part. Accel.
8. T. Garvey, S. Eylon, T.J. Fessenden and E. Henestroza, "Beam Acceleration Experiments on a Heavy Ion Linear Induction Accelerator (MBE-4)," to be published in Part. Accel.
9. T. Garvey, S. Eylon, T.J. Fessenden, K. Hahn, and E. Henestroza,"Transverse Emittance Studies of an Induction Accelerator of Heavy Ions," paper XRA51 1991 Particle Accelerator Conference, San Francisco, Ca.
10. S. Eylon, A. Faltens, W. Fawley, T. Garvey, K Hahn, E. Henestroza, and L. Smith, "Drift Compression experiments on MBE-4 and Related Emittance Growth Phenomena," paper XRA50 1991 Particle Accelerator Conference, San Francisco, Ca.

Beam Quality and Emittance in Free-Electron Lasers

C. W. Roberson

Physics Division
Office of Naval Research
Arlington, VA 22217

and

B. Hafizi*

Beam Physics Branch
Naval Research Laboratory
Washington, DC 20375-5000

Abstract

A retrospective look at the evolution of some of the ideas on the role of electron beam quality in the operation of a free-electron laser is presented. New results that include the effect of optical guiding are discussed and applied to the scaling of compact free-electron lasers.

* Permanent address: Icarus Research, Bethesda, MD 20814

In this paper we discuss the electron beam quality requirements for free electron lasers (FELs) and the unique demands this places on accelerators. We will review some of the ideas on this subject and discuss recent results when the radiation profile is determined by optical guiding.

At the 5th International Free Electron Laser Conference, with one exception [1], there was no discussion of beam brightness. There was, however, a formula, called the Lawson-Penner relation, which was widely employed in FEL research and which may be written in the form [2]

$$\varepsilon_n [\text{cm-mrad}] = 320 \sqrt{I[A]},$$

where ε_n is the normalized emittance in cm-mrad and I is the current in Amperes. The so-called Lawson-Penner relation was important since it connects two parameters, the emittance and the current, which are normally taken to be independent.

There were a number of papers in the literature showing the results from a wide variety of accelerators and the results seemed to follow this simple relationship over 5 orders of magnitude. The reasons for this were unclear.

The Lawson-Penner relation is based on the assumption that the brightness is the same for all accelerators. Although it is not surprising that this premise turned out to be false, it is remarkable that the relationship worked as well as it did. This appears to be due to a lack of development of new high current density cathodes and accelerating structures that preserve the beam quality. So, in a sense, the success of the Lawson-Penner relation is a testament to our complacency in high-brightness accelerator development. Interestingly enough, there is no evidence that Lawson or Penner ever discussed or published the relation, except to deny authoring it [3]. The scientific sociology that led to this is beyond the scope of this paper.

Among the notable exceptions to the Lawson-Penner relation, which stimulates some thought about the meaning and utility of brightness, are results from diodes immersed in a strong magnetic field [1,2]. In a variety of cases (with beam currents in the range 1-85 kA and current densities up to 400 kA/cm^2), θ, the ratio of the average perpendicular velocity to the parallel velocity, is about 30 mrad. What does this mean in terms of brightness? Brightness is defined as the current density per unit solid angle, Ω, about the axial direction; i.e.,

$$B = dI/(dA\ d\Omega),$$

which can be rewritten in terms of the current density, J, and θ as

$$B = J/(\pi\theta^2).$$

The brightness values for the field-free cases in Table I are approximately at the Lawson-Penner value. For the field-immersed cases, however, θ is approximately constant for current density varying over 2 orders of magnitude, which corresponds to 2 orders of magnitude improvement in brightness as compared to that in the Lawson-Penner relation. It is therefore natural to ask if a field-immersed diode may be used for FEL experiments? The answer is yes if the electron beam remains in the magnetic field throughout the interaction region. To consider why, let us examine brightness in terms of emittance. When the effect of space-charge is negligible, the (normalized) brightness can be written in terms of the normalized emittance ε_n as [2]

$$B_n = 2I/(\pi\varepsilon_n)^2.$$

For field-immersed diodes in which the beam is eventually extracted from the magnetic field the effective normalized emittance is

$$\varepsilon_{eff}^2 = \varepsilon_n^2 + (P_\theta/mc)^2,$$

where $P_\theta = |e|\Psi/2\pi$ is the canonical angular momentum, Ψ being the magnetic flux linking the cathode. Thus if it is intended to extract the beam from a field-immersed diode to be utilized in a field-free region--as is usually the case in FEL experiments--the presence of magnetic flux at the cathode dramatically increases the emittance of the beam. Consequently, it is necessary to have the magnetic field present in the FEL interaction region as well. That is, if one starts with a field-immersed cathode, it is necessary to have the field present all through the accelerator and the FEL.

It is also important to understand the relation of the beam brightness to the FEL interaction. The dynamics of the FEL is driven by the axial component of the velocity whereas the geometrical overlap (that is, the filling factor) is principally determined by the transverse components of the velocity. With this in mind, we can write the brightness as proportional to the quantity $J/(\delta\gamma_z/\gamma)$, where $\delta\gamma_z$ is the spread in the axial energy arising from emittance. In view of this it may be profitable to define a generalized beam quality in terms of an axial energy spread that includes all the contributions to the spread (e.g., that due to space charge) that is particularly suited to the FEL

$$B_Q = J/(\delta\gamma_z/\gamma).$$

So from the FEL point of view a good quality beam is one with a high current density and/or a small axial energy spread.

However, this is not sufficient since it is also necessary to ensure that the transverse profiles of the electron beam and the optical beam are well matched. For the simple case of a freely-diffracting Gaussian optical beam and an unfocused electron beam optimal geometrical overlap is achieved provided

$$\lambda = \pi\varepsilon.$$

For a FEL amplifier in the exponential gain regime of operation, the optical beam is tied to the electron beam as a result of gain focussing and it is necessary to generalize the expression $\lambda = \pi\varepsilon$. To this end, consider a planar wiggler with period λ_w and magnetic field B_w through which an electron beam with normalized emittance ε_n is propagating with velocity v. The radius for a matched electron beam is given by $r_b = (\varepsilon_n/2\gamma k_\beta)^{1/2}$, where $k_\beta = \sqrt{2}\pi a_w/\gamma\beta_z\lambda_w$ is the betatron wavenumber, γ is the relativistic factor, $a_w = |e|B_w\lambda_w/2\pi mc^2$ is the normalized wiggler vector potential and $\beta = v/c$.

In the high-gain Compton (exponential) regime of operation the radiation field maintains a constant spot size r_s through the wiggler as a result of gain focussing. Defining the filling factor by $f = (r_b/r_s)^2$, making use of the algebraic equation for f [4] and the expression for the matched electron beam radius, one obtains a relationship between the wavelength and emittance that includes the effects of electron beam focusing by the wiggler and optical guiding of the radiation beam [5]:

$$\lambda = \left\{ f_B \left[\frac{\gamma\nu}{2(1+a_w^2/2)} \right]^{1/2} \left[\frac{2+3f}{f(1+2f)^2} \right]^{3/2} \right\} \pi\varepsilon, \qquad (1)$$

where $\nu = I[kA]/17\beta_z$ is Budker's parameter, $f_B = J_0(\zeta) - J_1(\zeta)$ is the well-known difference of Bessel functions, with $\zeta = (a_w/2)^2/(1+a_w^2/2)$ and $\varepsilon = \varepsilon_n/\gamma\beta_z$ is the unnormalized electron beam emittance. Note that the wavelength of the radiation is, of course, given by the usual formula, $\lambda = \lambda_w(1+a_w^2/2)/2\gamma^2$; Eq. (1) expresses a scaling that permits us to relate the wavelength to the required electron beam quality as expressed by the transverse emittance.

Optimal coupling of the electron and the radiation beams is achieved for a filling factor f on the order of unity. Figure 1 is a plot of Eq. (1) for the two cases $f = 1/2$ and $f = 1$. The ordinate is chosen to be $\lambda/\pi\varepsilon$ so as to contrast our new scaling with the simple case $\lambda = \pi\varepsilon$.

Equation (1) may be employed in the design of a FEL. In Fig. 2 (a) we plot the beam voltage versus the normalized emittance for two different operating wavelengths, $\lambda = 5$ μm and $\lambda = 10$ μm, with a 200 A beam, a 1/2 kG wiggler field and a filling factor of 1/2. In connection with the two curves in Fig. 2(a) the following points must be emphasized. For a given voltage, operation at the shorter wavelength is more demanding in terms of the beam quality as expressed by the normalized emittance. For a fixed wavelength as the voltage is decreased from large values the emittance must be reduced too. However, below about 50 MV the scaling is seen to change. The transition in the two curves in Fig. 2 (a) comes about in the vicinity of the point where the normalized vector potential, a_w, equals unity. Figure 2(b), which shows the efficiency as a function of the voltage, also has a maximum around $a_w = 1$.

The extraction efficiency in Fig. 2(b) is given by [4]

$$\eta = \frac{f_B}{1 - 1/\gamma} \left[\frac{2f(2 + 3f)}{(1 + 2f)^4} \frac{\nu}{\gamma} \frac{a_w^2/2}{1 + a_w^2/2} \right]^{1/2}. \quad (2)$$

The expression for the efficiency given by Eq. (2) assumes that the electron beam is cold, with a negligible spread in the parallel velocity of the beam electrons. The predominant contributions to the spread arise from space-charge and beam emittance andd may be expressed as $(\delta\gamma_z/\gamma_z)_{sc} = \nu/\gamma$ and $(\delta\gamma_z/\gamma_z)_\varepsilon = (\varepsilon_n/r_b)^2/2(1 + a_w^2/2)$, and the extraction given by Eq. (2) is obtained provided

$$[(\delta\gamma_z/\gamma_z)^2_{sc} + (\delta\gamma_z/\gamma_z)^2_\varepsilon]^{1/2} < [(2 + 3f)/f]^{1/2}\eta. \qquad (3)$$

In Fig. 2 only the solid portions of the curves satisfy the constraint indicated by Eq. (2). Specifically, at low voltages the space-charge contribution to the the energy spread leads to the violation of the constraint in Eq. (3) and at high voltages the emittance contribution leads to its failure. Our purpose in displaying the curves in Fig. 2 over the extended range of emittance is to indicate the possible behavior if the constraint expressed by Eq. (3) were relaxed by some means.

Making use of the dispersion relation in the high-gain Compton regime for an electron beam with a Lorentzian or a Gaussian parallel velocity distribution, one can define an effective or scaled thermal velocity S [6]:

$$S \simeq \left(\frac{\delta v_z}{v_z}\right)\left(4\omega^2\gamma^3\gamma_z^2/\omega_b^2 a_w^2\right)^{1/3},$$

where δv_z is the spread in the axial velocity v_z, $\omega_b = (4\pi n_b e^2/m)^{1/2}$ is the plasma frequency, n_b is the beam density and $\gamma_z = (1 - v_z^2/c^2)^{-1/2}$. For $S > 1$ the phase velocity of the ponderomotive wave lies within the velocity spread in the beam frame. Using the requirements of good geometrical beam overlap ($f = 1$) and that the scaled thermal velocity must not exceed unity, one can use the expression for S to estimate the relative size of the parallel and perpendicular velocity spreads. Over a wide range of the beam voltage in Fig. 2 we find that the requirement on the parallel velocity spread is several orders of magnitude more severe than that on the perpendicular velocity spreads. This could have serious consequences for the efficiency of some FEL experiments.

Acknowledgement

This work was supported by the Office of Naval Research.

References

[1] C. W. Roberson, "Bright Electron Beams for Free-Electron Lasers," in *Free-Electron Generators of Coherent Radiation*, edited by C. A. Brau, S. F. Jacobs and M. O. Scully (SPIE, Bellingham, WA 1983) vol. 453, p. 25.

[2] C. W. Roberson and P. Sprangle, "A review of free-electron lasers," Phys. Fluids B, vol. 1, pp. 3-42, 1989.

[3] J. D. Lawson and S. Penner, "Note on the Lawson-Penner Limit," IEEE J. Quantum Electron. QE-21, p. 174 (1985).

[4] B. Hafizi, P. Sprangle, and A. Ting, "Optical gain, phase shift, and profile in free-electron lasers," Phys. Rev. A, vol. 36, pp. 1739-1746, 1987.

[5] C. W. Roberson and B. Hafizi, "Electron beam quality in free-electron lasers," Nucl. Instr. Meth. Phys. Res., vol. A 296, pp. 477-479, 1990.

[6] C. W. Roberson, Y. Y. Lau and H. P. Freund, "Emittance, brightness, free-electron laser beam quality, and the scaled thermal velocity," in *High Brightness Accelerators*, edited by A. K. Hyder, M. F. Rose and A. H. Guenther, New York: Plenum, pp. 627-645, 1988.

Figure Captions

Fig. 1 $(\lambda/\pi\varepsilon)$ versus $f_b[\gamma\nu/2(1 + a_w^2/2)]^{1/2}$ for filling factor $f = 1/2$ and $f = 1$.

Fig. 2   (a) Beam voltage versus normalized emittance; (b) extraction efficiency versus beam voltage. Solid part of curves corresponds to cold-beam interaction.

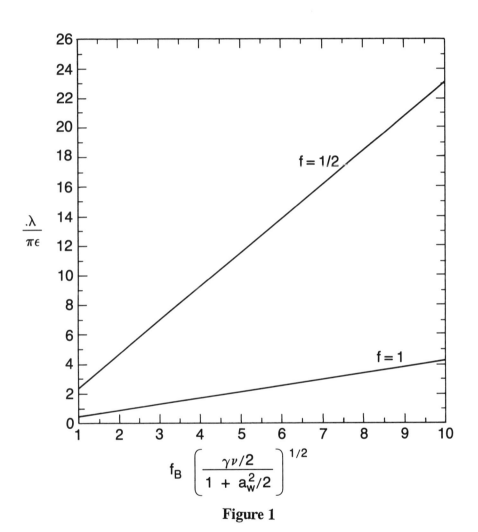

Figure 1

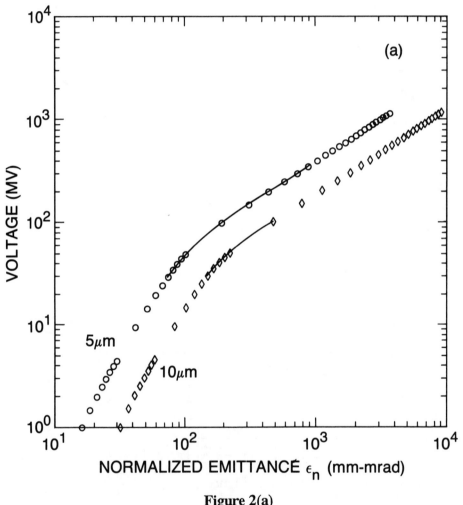

Figure 2(a)

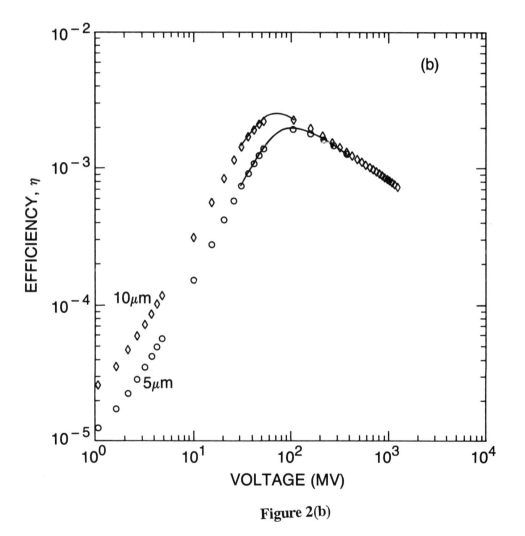

Figure 2(b)

The Los Alamos High-Brightness Photoinjector

Patrick G. O'Shea
MS J579, Los Alamos National Laboratory, Los Alamos NM 87545

ABSTRACT

For a number of years Los Alamos National Laboratory has been developing photocathode RF guns for high-brightness electron beam applications such as free-electron lasers (FELs). Previously thermionic high-voltage DC guns have been the source of choice for the electron accelerators used to drive FELs. The performance of such FELs is severely limited by the emittance growth produced by the bunching process and also by the low peak current of the source. In a photoinjector, a laser driven photocathode is placed directly in a high-gradient RF accelerating cavity. A photocathode allows unsurpassed control over the current, and the spatial and temporal profile of the beam. In addition the "electrodeless emission" avoids many of the difficulties associated with multi-electrode guns, i.e. the electrons are accelerated very rapidly to relativistic energies, and there are no electrodes to distort the accelerating fields. For the past two years we have been integrating a photocathode into our existing FEL facility by replacing our thermionic gun and subharmonic bunchers with a high-gradient 1.3 GHz photoinjector. The photoinjector, which is approximately 0.6 m in length, produces 6 MeV, 300 A, 15 ps long electron micropulses at a 22 MHz rep. rate. The beam is then injected into an RF linac, and accelerated to a final energy of 40 MeV. We have recently begun lasing at wavelengths near 3 μm.

INTRODUCTION

The efficient conversion of electron energy into photon energy requires low emittance, high-brightness beams[1]. Free-electron laser oscillators operating at high power and short wavelength (λ_o) require high-current, low-emittance (ε_n) electron beams. The gain of an FEL increases with beam current subject to the constraint that

$$\lambda_o > \frac{\varepsilon_n}{4\pi\beta\gamma} \tag{1}$$

where $\varepsilon_n = 4\pi\beta\gamma(<x^2><x'^2> - <x.x'>^2)^{1/2}$ is the normalized emittance. In this context high current implies I»100A and low emittance implies ε_n « 100 π mm-mrad . Very low emittance allows the possibility of accessing short optical wavelengths at low beam energy, by lasing on harmonics of the fundamental FEL wavelength. The FEL resonance condition is given by:

$$\lambda_0 = \frac{\lambda_w}{2N\gamma^2}(1 + K^2) \qquad (2)$$

where: K is the wiggler parameter typically of order 1, γ is the relativistic energy parameter, λ_w is the wiggler period, N is the harmonic number with N = 1,3,5... If the emittance satisfies the constraint of eqn. 1, then harmonic lasing may be possible at short wavelengths. The electron beam energy required to produce a given wavelength scales as $\frac{1}{\sqrt{N}}$, and produces a corresponding saving in the size and cost of the accelerator required.

There are two issues of importance here. The first is the intrinsic brightness of the source and the second is the ability to transport the beam with minimal emittance growth through the linac.

Previously high brightness beams have been produced by thermionic high-voltage guns, with emittances near the source thermal limit. We can define the normalized emittance for a source in terms of the cathode radius (r_c) and temperature (T)[2]:

$$\varepsilon_n = 2\pi r_c \sqrt{\frac{kT}{mc^2}} \qquad (3)$$

and the normalized brightness in terms of the current density J:

$$B_n = \frac{2I}{[\varepsilon_n]^2} = \frac{mc^2 J}{2\pi kT} \qquad (4)$$

Table 1 shows a comparison between the best thermionic cathode and photocathode sources for RF linacs[3].

Table 1. Comparison of thermionic cathode and photocathode

Cathode	**Thermionic**	**Photocathode**
T: Effective temp. (eV)	0.1	0.2
J: Current density (A/cm^2)	10	600
Brightness (A/(m-rad)^2	1 x 10^{11}	2 x 10^{12}

Photocathode has intrinsically higher brightness and reduced emittance growth in transport through the linac. The design of an integrated photoinjector linac removes the requirement for subharmonic RF bunching at non-relativistic energies and of magnetic bunching common in linacs with high-voltage thermionic sources, thereby removing many of the emittance growth opportunities. For a review of RF guns in general see reference 4.

For a number of years we have been developing photocathode RF guns for high-brightness electron beam applications[3]. In a photoinjector, a laser driven photocathode is placed directly in a high-gradient RF accelerating cavity (see fig. 1). This system allows unsurpassed control over the spatial and temporal profiles, and current of the beam. In addition the "electrodeless emission" avoids many of the difficulties associated with multi-electrode guns,i.e. the electrons are accelerated very rapidly to relativistic energies, and there are no electrodes to distort the accelerating fields.

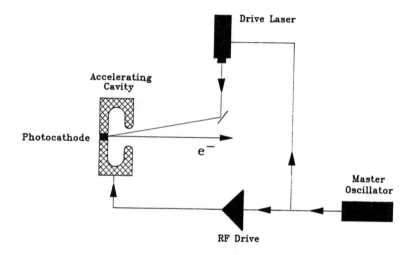

Fig 1 Photoinjector schematic

Drive Laser and Photocathode

There have been a number of approaches proposed for the photocathode/drive laser combination necessary for a photoinjector. These approaches are divided into two classes: Low quantum efficiency and high quantum efficiency. (Quantum efficiency, Q, is defined as the number of free electrons produced off the cathode per incident photon). Low Quantum efficiency cathodes ($Q < 0.1\%$) have typically the following characteristics[5-7]:

    metal or LaB_6 cathode

    long cathode lifetime

    very high drive laser power required in ultra-violet (e.g.quadrupled Nd:YAG)

    not readily scalable to high repetition rate or duty factor.

High Quantum Efficiency Cathodes (Q > 1%) are characterized by[3]:
   multi-alkali(e.g. CsK_2Sb)
   short cathode lifetimes (days)
   high Q in the visible
   scalable to high duty factor

Following from the definition of quantum efficiency above, we can readily obtain the fundamental relationship between drive-laser wavelength (λ_{DL}), and pulse energy (**E**), photocathode quantum efficiency (**Q**) and charge (**C**) per pulse is given by:

$$C = \left[\frac{e}{hc}\right] QE\lambda_{DL} \qquad (5)$$

where e = electronic mass, h = Plank's constant, and c = velocity of light.
An alternative formulation of this equation is in terms of current (**I**) from the cathode and drive laser power (**P**):

$$I = \left[\frac{e}{hc}\right] QP\lambda_{DL} \qquad (6)$$

Equation (6) is only valid for the case of prompt electron emission from the cathode i.e. the emission time is much less than the drive laser pulse length. Equation (6) may be rewritten in practical units:

$$I(A) = 8.08 \times Q(\%) \times P(kW) \times \lambda_{DL}(\mu m) \qquad (7)$$

In the case of the Los Alamos experiment **I** = 0.1 A (macropulse average), **Q** = 6% and λ_{DL} = 0.527 µm (doubled Nd:YLF), then **P** = 4 W. For a low quantum efficiency cathode **Q** < 0.01%, at λ_{DL} = 0.26 µm (quadrupled Nd:YAG), then **P** = 4800 W for the same 0.1 A macropulse average current. Note that this is the drive laser power actually delivered to the cathode. The low quantum efficiency cathodes do not appear practical for high average current applications because of the high drive laser power required.

At Los Alamos we have chosen a high **Q** material CsK_2Sb. Typically **Q** is greater than 6% at the start of an accelerator run. However such cathodes are sensitive to contamination in poor vacuum. Typically the vacuum in the operating photocathode cell is less than 1 x 10^{-9} torr. The 1/e lifetime for decay of **Q** is greater than 10 hrs when the accelerator is operating. Our design requirement is to produce 5 nC per 10 ps micropulse at the cathode, in which case we need to maintain our **Q** x **E** product greater than 1.1 %µJ for effective operation. Since our drive laser can deliver up to 6 µJ per micropulse to the cathode we can operate with **Q** as low as 0.2 %.

The performance of our FEL depends critically on our photocathode and its drive laser. Phase and amplitude jitter in the drive laser result in energy and current jitter in the electron beam. We require the phase and amplitude jitter to be <1 ps and <1 % respectively. Table 2 gives the measured performance of the drive laser and photocathode.

Table 2 Drive laser and photocathode performance

Drive laser:	Doubled Nd-YLF
Wavelength	527 nm
Micropulse width	7-15 ps
Micropulse rep. rate	21.7 MHz
Micropulse energy	12 µJ (5-6 µJ at cathode)
Macropulse length	0-200 µs
Phase jitter	< 1 ps
Amplitude jitter	< 1%
Photocathode:	CsK_2Sb
Radius	4 -5 mm
Peak quantum efficiency	8 %
1/e lifetime in operating accelerator	10-15 hrs at 2×10^{-9} Torr

To improve our operating time on a single cathode we are endeavoring to a) reduce the quantum efficiency decay rate; and b) increase the energy delivered by the drive laser to the photocathode, c) install a multi-cathode transfer system.

Reducing the decay rate of **Q** implies improving the vacuum conditions in the accelerator. Studies have shown that CO_2 and H_2O can contaminate the cathode and reduce its effective lifetime[8]. Since a standard bake at 250-350°C imparts « 1 eV to surface adsorbed molecules, it is not effective in removing those adsorbed gases that are bound with binding energies » 1 eV. In the non-operating accelerator (no RF, no beam) such a bake produces a vacuum of 5×10^{-10} torr. In the operating accelerator there are many electrons with energies » 1 eV that induce electron stimulated desorption (ESD) of gases from the cavity walls and cause the pressure to rise to the mid 10^{-9} torr range. To improve this situation we have initiated an RF generated glow discharge cleaning technique[9]. Using 200 W CW 1.3 GHz RF, fed into the photoinjector cavity through the waveguide, we generated a glow discharge with approximately 10^{-2} torr of hydrogen. By varying the RF frequency (1.3 ±0.03 GHz) the glow could be initiated in one or more cells of the photoinjector. During the discharge the photoinjector was maintained at a temperature of 130 °C. The discharge was run for 48 hours, followed by a 24 hour bake at only 130 °C. The immediate result was to reduce the pressure to 1×10^{-10} torr at room temperature. Preliminary results indicate that the pressure during high power RF operation has been reduced to the low 10^{-9} torr range.

We have implemented a drive laser upgrade which has increased the deliverable optical energy to the photocathode from 1 to 6 μJ per micropulse.

At present we have a single cathode which must be refurbished when its **Q** falls below allowable limits. This refurbishment process involves retraction of the cathode into a preparation chamber and the bake-off of the residual cathode material that is only a few tens of nanometers thick, followed by the redeposition of the cathode material. This process takes about 3 hours. So as to reduce this downtime we plan to replace the single cathode system with a six-cathode casette. These cathodes will be fabricated away from the accelerator and brought to the accelerator in a portable vacuum chamber. When this system is in place we will not need a preparation chamber attached to the accelerator. The accelerator runtime between casette changes should be in excess of a week.

The APEX Photoinjector

We have installed and tested a high-gradient (26 MV/m at the cathode) 1300 MHz, $\pi/2$-mode photoinjector, that is 0.6 m long and produces 6 MeV, 300 A, 15 ps electron pulses at a 22 MHz rep. rate[10]. The photoinjector includes a focusing solenoid following the prescription of Carlsten for emittance compensation[11]. Figure 2 shows a cutaway view of the actual Los Alamos 6 MeV photoinjector. Table 3 gives the specifications for the photoinjector. The photocathode cell is a half cell, followed by five full accelerating cells. The RF feed is through the sixth cell. The accelerating cells are coupled by on axis coupling cells. The RF structure is enclosed in a stainless steel vacuum jacket, to which are attached a titanium sublimation pump and an ion pump. Pumping from the accelerating cells into the vacuum manifold is achieved through 16 longitudinal slots per cell, giving an effective pumping speed of approximately 170 l/s per cell. The resultant pressure at the photocathode is typically less than 10^{-9} torr. The photocathode itself is on a molybdenum plug that is attached to an actuator. The cathode can be retracted from the accelerator into the photocathode preparation chamber.

Following the photoinjector electron beam is accelerated to 40 MeV by three additional side-coupled linac tanks. RF power is provided by Thomson CSF klystrons (TH2095A), with one klystron per accelerator tank.

The FEL configuration is a single-accelerator master-oscillator power-amplifier (SAMOPA)[12-14] configuration. Resonator optics are often the limiting factor in high average power FELs. In the SAMOPA concept the electrons first pass through a low power oscillator and then through a high gain amplifier. The light from the oscillator is fed into the amplifier. Since the power in the oscillator is low, and there are no resonator optics in the amplifier, the optical damage difficulty is removed. We are studying the physics issues associated with SAMOPA operation as part of the Boeing/Los Alamos collaboration to build the Average Power Laser Experiment (APLE). Los Alamos will perform the APLE prototype experiments and will be known by the acronym APEX.

Figure 2. The APEX photoinjector

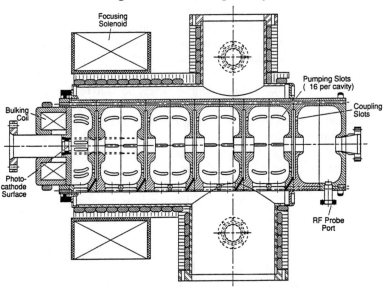

Table 3 APEX Photoinjector specifications

Frequency	1300 MHz
Accelerating gradients:	
cell 1	26.0 MV/m
cell 2	14.4 MV/m
cell 3-6	10.0 MV/m
Measured Q	18500
Shunt impedance	35 MΩ/m
Copper power	1.8 MW
Output energy	6 MeV
Micropulse length	15 ps
Micro pulse charge	5nC
Micropulse rep. rate	21.7 MHz
Peak current	300 A
Macropulse length	100 µs
Macropulse rep. rate	1 Hz
Macropulse ave. current	0.1 A
Emittance (4rms, normalized)	< 50 π mm-mrad

In 1990 we completed experiments that characterized the photoinjector operation and beam transport through one additional accelerator tank at an energy of 15 MeV.

RF Controls

The stability of the RF phase and amplitude is as critical to the FEL performance as is that of the drive laser. We have replaced our old RF feedback control system with a novel system using state-feedback[15]. The system in its present form is significantly smaller and produces better RF stability than our old system. Table 3 gives the performance of the state-feedback system over a 100-μs macropulse on the photoinjector.

Table 4 RF phase and amplitude stability

	Amplitude (%)	Phase (ps)
Jitter	0.03	0.1
Slew	0.25	1

We have recently testing the effectiveness of the feedback system on all four accelerator tanks.

Operational Experience

Measurements on the electron beam produced by the photoinjector have been made after post acceleration to 14 MeV by an additional side coupled tank. Of particular interest has been the comparison between the design code (INEX) predictions and actual performance. Details of the comparison between INEX and measurements have been presented elsewhere[16].

The performance of the photoinjector has proven to be excellent in the areas of most importance to FEL operation, i.e. reduced emittance and reduced energy spread and increased brightness as indicated in table 5.

Table 5 Comparison of the performance of the old vs. new injector at the Los Alamos FEL

Electron source	Thermionic gun	Photoinjector
Emittance	160 π mm-mrad	40 π mm-mrad
Energy spread	0.5%	0.3%
Charge per bunch	5 nC	5 nC
Peak Current	300 A	300 A
Brightness	2.4×10^9 A/(m-rad)^2	4×10^{10} A/(m-rad)^2

There were three unanticipated effects observed during operation of the photoinjector:

1) Multipactoring in one or more coupling cells produced coherent 7 MHz oscillations in both phase (±1°) and amplitude (±1%) of the RF in the tank. The problem was solved by detuning the photocathode cell (the end wall was pulled by a couple of tenths of mm) so as to raise the fields in the coupling cells above the multipactoring limit.

2) The electron beam was observed to have an elliptical crossection before passing through any quadrupole magnets. The source of this effect was RF quadrupole focusing resulting from the number and location of the coupling slots in the accelerator. This has caused an emittance growth of approximately 10 π mm-mrad. This problem is being partially corrected by placing a small quadrupole magnet between the photoinjector and the next tank. Future photoinjector designs will have a modified coupling slot arrangement which will eliminate this emittance growth problem. For more details see ref. 16.

3) A small field-emission electron current (0.1 mA) was observed with the drive laser off. The intensity of this field emission current is not sufficient to significantly affect our operation. For more details see ref. 17

Present Status

The commissioning of the complete 40 MeV linac with the integrated photoinjector is complete. On June 21, 1991 first lasing was observed at a wavelength of 3.5 µm with an extraction efficiency of 0.6% over a 40 µs macropulse. We are continuing to characterize the accelerator and FEL performance. Details of the designed performance of the FEL system are given in ref. 11.

Conclusion

We have shown in practice that photocathodes integrated into RF linac in the form of photoinjectors are a viable and practical alternative to thermionic sources. In the Los Alamos case replacing the thermionic gun and subharmionic bunchers with a photoinjector has increased the electron beam brightness by a factor of 20. We believe that high quantum efficiency cathodes are the only practical choice for high average current electron accelerators because the drive laser power is prohibitively high for low quantum efficiency sources. The present disadvantage of high quantum efficiency sources is the relatively short cathode lifetime. We have made significant progress in understanding the cathode decay processes and in improving the lifetime. We continue to work on more advanced cathode designs that will lead us to both high quantum efficiency and long lifetimes. Based on our experience with the APEX photoinjector we are constructing new 20 MeV photoinjector as part of the Advanced FEL (AFEL) project. The electron beam from the AFEL accelerator should be an order of magnitude brighter than that of APEX. In addition we will use the experience gained on APEX to design and build the high average power APLE free-electron laser.

7. Acknowledgements

This work was supported and funded by the US Department of Defense, Strategic Defense Initiative Organization and Army Strategic Defense Command, under the auspices of the US Department of Energy. The successful operation of the APEX free-electron laser was made possible through the efforts of a large number of dedicated people. The following deserve special recognition for outstanding contributions to the project: K. McKenna and R. Sheffield for overall project guidance; D. Feldman for accelerator and FEL commissioning; B. Carlsten, J. Goldstein and M. Schmitt and B. McVey for theoretical support; D. Schrage and L. Young for the design and construction of the photoinjector; N. Okamoto, P. Schafstall, R. Springer and S. Volz for the photocathode subsystem;G. Busch, J. Barton and J. Early for the photocathode drive laser, L. Connor, J. Johnson, N. Okay, P. Ortega, W. Stein and T. Zaugg for the RF system, S. Apgar, R. Feldman and A. Lumpkin for electron beam diagnostics and data analysis; S. Bender, D. Byrd, M. Feind N. Okamoto for FEL resonator and optics; R. Martinez and R. Stockley for mechanical support

8. References

1. For an introduction to FEL physics see "Free-Electron Lasers" by C.A. Brau, in Advances in Electronics and Electron Physics, Suppliment 22,Academic Press (San Diego), (1990).

2. For discussion of emittance and brightness see "The Physics of Charge-Particle Beams" 2nd ed. by J.D. Lawson, Clarendon Press (Oxford), (1988)

3. R.L Sheffield, "Photocathode RF Guns", in Physics of Particle Accelerators, vol 2, AIP Conference Proc. #184, pgs 1500-1531.

4. C. Travier, "RF Guns: Bright Injectors for FEL, Nucl. Inst. Meth. **A304,** 285, (1991)

5. M. Curtin, G. Bennett, R. Burke, A. Bowmik, P. Metty, S Benson, J.M.J. Madey, Nucl. Inst. Meth, **A296,**127, (1990)

6. S. Hartman et. al to appear in the Proc. 1991 IEEE Particle accelerator Conf.

7. K. Batchelor, Proc. 1990 Linear Accelerator Conf., p. 810, Los Alamos publication LA-12004-C (1991)

8. R.L. Sheffield, Proc. 1990 LINAC Conf., p. 269, Los Alamos Pub. # LA-12004-C (1991)

9. "Surface Conditioning of Vacuum Systems", R. Langley Ed., American Vacuum Soc. Series, Vol. 8, AIP, (1990)

10. L.M. Young, Nucl. Inst. Meth., **B56/57**, 978, (1991)

11. B. E. Carlsten, R. L. Sheffield, Proc. 1988 Linear Accelerator Conf. CEBAF report #89-001, 365, (1989)

12. D.W Feldman, W.D. Cornelius, S.C. Bender, B.E. Carlsten, P.G. O'Shea, R.L Sheffield, Free-Electron Lasers and Applications, D. Prosnitz, Ed., Proc. SPIE **1227**, 2, (1990)

13. B.E. Carlsten, L.M. Young, M.E. Jones, B. Blind, E.M. Svaton, K.C.D. Chan, L.E. Thode, Nucl. Inst. Meth., **A296**, 687, (1990)

14. J.C. Goldstein, B.E. Carlsten, B.V. McVey, Nucl. Inst. Meth., **A296**, 273, (1990)

15. W.J.D. Johnson, C.T. Addallah, Proc. 1990 LINAC Conf. page 487 (1991)

16. B.E. Carlsten, et al, to appear in IEEE J. Quantum Electronics (June) (1991)

17. A.H. Lumpkin, "Observations on Field-Emission Electrons from the Los Alamos FEL Photoinjector", Proceedings of the 1991 Particle Accelerator Conference, to be published

H⁻ ION SOURCES*

J.G. Alessi
AGS Department, Brookhaven National Laboratory
Upton, New York 11973

ABSTRACT

A review is given of H⁻ ion sources, with the emphasis on sources of use for accelerator applications. A brief description is given of magnetron, multicusp/converter, Penning, and volume H⁻ ion sources. Operating parameters for examples of the various type sources are presented, and then some comparisons among the sources are made.

I. INTRODUCTION

There is an ongoing interest in H⁻ ion source development, due to uses of these sources in several areas. In high energy proton accelerators, H⁻ sources are used due to the efficiency of charge exchange injection into synchrotrons or cyclotrons. In magnetic confinement fusion, the interest in H⁻ sources comes from the need for high energy neutral beams for plasma heating and current drive. There, one starts with negative ions rather than H⁺ due to their higher neutralization efficiency. Defense applications also desire intense neutral beams for particle beam weapons, so again, one starts with negative ions to take advantage of the higher neutralization efficiency. While the exact source requirements for the three areas are quite different, there is a large overlap in the source development for the three areas with respect to attempting to understand the fundamental processes in the various H⁻ sources, optimization of source output, etc.

In this paper the main types of H⁻ sources presently in use will be discussed. The emphasis will be on operating parameters of the various sources, and comparisons of features which are generally of importance when trying to select a source for a particular application. The performance of multiampere H⁻ sources being developed for fusion will not be included, although their basic features will be the same as the smaller sources that are mentioned. More details on specific sources, as well as discussions of the fundamental processes in these sources, can be found in the Proceedings of the International Symposium on the Production and Neutralization of Negative Ions and Beams, held every three years at BNL. The proceedings of the most recent two (1986 and 1989) are given as Refs. 1,2.

---

*Work performed under the auspices of the U.S. Department of Energy.

II. SOME CRITERIA FOR H⁻ SOURCE COMPARISONS

Before discussing specific sources, some aspects of source performance that should be considered when making comparisons will be mentioned. One normally thinks first of choosing the source that can deliver the required current in the smallest emittance. However, as discussed below, one must also consider parameters such as the operating pressure of the source, gas flow from the source, power efficiency, cesium consumption, etc.

Emittance
The definition for normalized emittance is:

$$\epsilon_{n,f} = \beta\gamma \, A_f/\pi \tag{1}$$

where A_f is the marginal phase space area for a specified beam fraction f. The normalized rms emittance definition used here is:

$$\epsilon_{n,rms} = \beta\gamma[\overline{x^2}\,\overline{x'^2} - (\overline{xx'})^2]^{\frac{1}{2}} \tag{2}$$

Care should be taken when rms emittance is given in other references, in that the definition sometimes includes an additional factor of 4.

In some cases, the measured rms emittance of the full beam is calculated directly from eq. 2. This sometimes presents problems in determining where the beam actually ends relative to a background of "noise" from the measurement electronics.

Experience shows that a Gaussian distribution normally best describes real beams, in which case the emittance for a beam fraction f is found to be related to that fraction by[3]:

$$f = 1-e^{-(\epsilon_f/2\epsilon_{rms})} \tag{3}$$

If one plots the measured fractional emittance of a beam, ϵ_f, vs. $\ln[1/(1-f)]$, one normally sees a linear region at low beam fraction where a Gaussian distribution is a good description of the beam, and a deviation from this linear dependence at higher beam fractions, where beam halo, extractor aberrations, etc. cause a deviation from Gaussian. Frequently, rather than using eq. 2 directly, the rms emittance quoted for an ion source will be that obtained from the slope of this ϵ_f vs. $\ln[1/(1-f)]$ curve at lower beam fractions.

If one assumes that the ions are produced with a Maxwellian energy distribution with temperature kT in the ion source, it can be shown that the rms emittance is given by:[4]

$$\epsilon_{n,rms} = \tfrac{1}{2} r \, (kT/Mc^2)^{\tfrac{1}{2}} \qquad (4)$$

for a circular source anode aperture of radius r. This allows one to extract an "effective" ion temperature, kT, for a source from the rms emittance obtained at low beam fractions. It is "effective" since things such as emittance growth from plasma noise are still included.

The method of determining the rms emittance and ion temperature from eqs. 3 and 4 is useful when trying to understand the processes occurring in the source, since the influence of the extraction optics on the beam is reduced. However, people in the accelerator community frequently quote instead the total emittance of the beam from the source at a 90% or 95% beam fraction. In this case, this is the "bottom line", the emittance that must be transported and accelerated.

In what follows, to compare sources with emittances quoted in different ways, eq. 3 will be used to convert ϵ_f to ϵ_{rms} (ex. ϵ_{rms}=0.22 x $\epsilon_{90\%}$; 0.167 x $\epsilon_{95\%}$). Unfortunately, this will often give a larger value for the rms emittance than the second method, since this assumes that the beam distribution is Gaussian out to these high beam fractions, which it frequently is not.

Brightness

The normalized rms beam brightness is defined as:

$$B_{n,rms} = 2I/[\pi^2 \, \epsilon_{n,rms}(x) \, \epsilon_{n,rms}(y)] \qquad (5)$$

where I is the beam current. It is interesting to substitute the emittance expression from eq. 4 into eq. 5. Neglecting constants, one obtains:

$$B_{n,rms} \propto J/kT \qquad (6)$$

where J is the current density of the extracted beam at the anode aperture. Therefore, for a high brightness beam, one desires not only a low ion temperature in the source, but a high extraction current density. These are frequently conflicting requirements (i.e. as the arc is increased to bring up the current density, the ion temperature may also increase). One might expect from eq. 6 that the brightness of a source would be constant, independent of the total extracted current, since the current could be increased by merely increasing the aperture size and keeping the current density constant. However, in reality an increase in aperture often results in a reduced current density due to pressure limitations, increased gas flow out of the source causing increased stripping, poorer extraction optics, etc., and the brightness usually decreases with increasing current.

Relevance of Some Other Source Parameters

From a practical point of view, other properties of a source should also be considered. The gas flow from a source is particularly important since it determines pumping requirements, and also because it directly effects high voltage holding and H⁻ beam losses due to stripping. For a given beam current, the gas flow will be proportional to the operating pressure in the source, but approximately inversely proportional to the beam current density (smaller aperture required).

The size of a source (discharge chamber volume) can effect performance in ways beyond plasma/H⁻ ion production effects. At high duty factors, one typically needs a larger source in order to keep the arc power density reasonable. At low duty factors, however, a smaller source is often better because one can more effectively pulse the gas, reducing the average gas flow out of the source. For example, in a magnetron H⁻ source, having a source volume of only ≈ 1 cm^3, the gas can be pulsed effectively even at many 10's of Hz. A volume H⁻ source, however, typically has a discharge volume of hundreds to thousands of cm^3, making it senseless to try pulsing unless the rep rate is very low.

The ratio of extracted electrons to H⁻ ions is important in terms of high voltage power supply loading, difficulty in voltage holding and dumping the electrons in a controlled manner, and the influence on extraction optics. Finally, cesium consumption, source filament lifetime, and arc power density can influence the reliability of the ion source. From the above, one can see that there is still some "art", as well as (hopefully) science, to the development of the best source for a particular application.

III. MAGNETRON SURFACE PLASMA SOURCE

The magnetron H⁻ source has a coaxial electrode geometry, with the cathode the inner and the anode the outer electrode. Electrons perform an ExB drift around the cathode due to the presence of a transverse magnetic field in the discharge region, and ionize the high pressure hydrogen gas injected into this region. A dense hydrogen plasma is thus produced in the discharge between the cathode and anode. Cesium is added, lowering the work function of the cathode as well as enhancing the discharge (reducing the arc impedance). Plasma ions (H⁺ and Cs⁺) are accelerated to the cathode, where H⁻ ions are produced via backscattering or desorption from the low work function cathode. The ions are then accelerated in the cathode sheath, cross the plasma, and are extracted through a hole or slit in the anode. The transverse ion energy from surface production results in an effective ion temperature determined from the beam emittance of ten's of eV. In addition to the ions leaving the source with ≈ 150 eV from the cathode, slow ions are produced by charge exchange of the primary ions with background H between the cathode and anode. Therefore, two energy components can be observed in the output beam.[5]

This type source is compact, with only a few mm gap between the cathode and anode. The source pressure is in the 100 mT range, however since the extracted H⁻ current density can be several A/cm^2 (and gas is typically pulsed), the gas flow is not a problem. The extracted electron current is often less than the H⁻ current. There is a lot of operational experience with this source on high energy accelerators, where it routinely gives 3-6 months of continuous operation in the pulsed mode. The source is used at BNL,[6] FNAL,[7] ANL,[8] and DESY,[9] and has also been operated on a test stand at SSC. The BNL source requires only about a 10-15 A arc current to produce 65-100 mA of H⁻ (with similar performance at the other labs), and this very good arc efficiency is part of the reason for its reliability. A variation of this source developed in the USSR, the semiplanotron, has produced H⁻ currents of up to 11 A (600 apertures).[10] Operating parameters of several magnetron sources are given in Table 1.

IV. MULTICUSP / CONVERTER H⁻ SOURCE

This is also a surface-plasma H⁻ source, with the H⁻ ions again produced on a low work function (cesium coated) converter surface. Plasma generation is in a large volume ("bucket source"), via a discharge between the hot filament cathode and the chamber walls. The use of a filament allows operation at a much lower pressure than in the magnetron. The plasma density is also much lower than in the magnetron, however, resulting in a much lower extracted H⁻ current density. Ions produced on the negatively biased converter are accelerated away from the converter and will leave the source without any extraction field. This allows one to have the accelerating voltage removed from the plasma region, resulting in very few extracted electrons. While the transverse energy of the surface produced H⁻ is similar to that in the magnetron source, one can limit the extracted ions to those below a certain transverse energy by adjusting the converter and anode aperture geometry, allowing one in a sense to "select", at the cost of arc efficiency, the effective ion temperature, which is a few eV in operating sources. The large discharge volume and low plasma density allows the source to operate easily steady state, making this the source of choice at this time for accelerator labs requiring both high duty factor and high currents. In spite of the low source pressure, the gas flow is quite high because a large aperture is required due to the low current density. This large aperture also leads to a higher Cs consumption than the magnetron. The source is used at LAMPF,[11] where it can operate for ≈40 days continuously. At KEK,[12] lifetimes are ≈3 months due to the use of a LaB_6 filament. More than 1 A of H⁻ has also been produced from this source steady state at LBL.[13] Table 1 gives some parameters of this type source.

V. PENNING SOURCE

The Penning source also relies on surface production of H⁻ ions on a low work function surface. However, in this case there is no direct path for the surface produced H⁻ ions to reach the extraction aperture. Rather, these fast surface produced ions undergo charge exchange with background H⁰ in the discharge, and the slow H⁻ ions are extracted from the plasma. This results in an H⁻ ion temperature of less than 1 eV.[14] There is also the possibility that some volume production of H⁻ is occurring in this source (see section VI.).

The ion source is small, with a high plasma density making it capable, like the magnetron, of high current densities. This high current density, coupled with the low ion temperature, results in the highest brightness beam of all the H⁻ sources. However, in order to run with the "quiet" discharge required to preserve this small emittance, the source pressure and cesium consumption are higher than that of the magnetron. The arc efficiency of this source is not as good as that of the magnetron since one is not directly extracting the surface produced ions. The small size, high current density, and reduced arc efficiency makes reliable high duty factor operation at high currents difficult. In defense applications, where brightness is a very important parameter, this has been the source of choice. The most active development work on this source takes place at Los Alamos.[15]. The only high energy accelerator use of the Penning source is at Rutherford Lab,[16] where emittances measured after acceleration to 665 keV in a Cockcroft-Walton are higher than those measured at a similar point at FNAL, where a magnetron is used. Although this may partly be due to differences in the optics, another contributing factor is probably the difficulty in operating reliably for long periods in the "quiet" discharge mode. Table 1 gives parameters of several Penning sources.

VI. VOLUME PRODUCTION H⁻ SOURCE

This type H^- source is the most recently developed, and is presently receiving the most attention. H^- ions are produced directly in the plasma volume of the source, without the need for cesium. The primary H^- production mechanism is believed to be dissociative attachment of vibrationally excited H_2, $e + H_2(\nu) \rightarrow H^- + H$. The source is usually divided into two regions. In the main discharge region, there is ionization by high energy electrons for plasma production, and vibrationally excited hydrogen molecules are produced by fast electron impact, wall processes, etc. A magnetic "filter field" separates this plasma production region from the extraction region. Plasma ions and neutral species pass through the filter field. Slow electrons, which are required for the dissociative attachment, can also readily diffuse across this filter field. However, the crossing by fast electrons, which contribute to the destruction of H^- ions, is impeded. This produces a region which is more favorable for H^- production.

There are still many unanswered questions concerning the operation of these sources, such as the mechanism by which the vibrationally excited molecules are produced, wall effects in the source, and the effect of filter field shape. Because this source is still in the developmental stage and there is only very limited operational experience, this type source is the most difficult to summarize. Typically, the chamber volume is large (hundreds of milliliters to several liters). The H^- ion temperature can be < 1 eV, but the H^- current density is usually low, so the beam brightness is so far only average. The source pressure is typically 10-30 mT, so with the low current density, gas flow from the source can be quite high. Because the H^- is extracted directly from the plasma, there is a tendency to extract a large amount of unwanted electrons. The electron-to-H^- ratio can be several hundred, although the volume source being developed at BNL has an e^-/H^- ratio of 2-5 for currents of up to 30 mA.[17]

Volume sources typically use a hot filament as the cathode for the discharge. This raises questions about the lifetime of the source for accelerator applications. Recently, however, a volume source has been operated with an rf antenna in the source for plasma production.[18] This works as well or better than filaments, and promises a better source lifetime. It has also been found that the addition of cesium into the volume source enhances the H^- output by factors of ≈2-10, while at the same time reducing the extracted electron current and the operating pressure.[19] Among the contributing factors in the improved performance with Cs are probably that some surface production of H⁻ is occurring, and that the Cs is increasing the supply of cold electrons, desirable for the dissociative attachment process. The cesium consumption is not yet quantified. Barium has also been found to enhance the H⁻ output.[20] A large volume source at JAERI has produced currents of up to 10 A in 1 second pulses with the addition of cesium.[21]

Examples of volume source performance are given in Table 2. While operational experience is still very limited, this type source has the potential of fulfilling the needs of almost all H⁻ source users.

VII. SUMMARY

The sources shown in Tables 1 and 2 are meant to be representative of the various types, rather than all inclusive. Generally, only the performance of smaller, single aperture versions of a source are given, and usually only cases where the emittance of the source was also measured. The rms brightness of various sources is plotted as a function of the H⁻ current in Fig. 1. (All points are taken from the Tables). Finally, I have attempted to make some generalizations concerning the performance of the various type sources in Table 3.

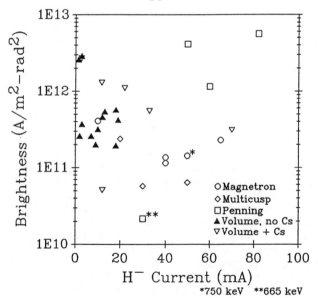

Fig. 1. Normalized, rms brightness of various sources vs. H⁻ current.

TABLE 1 - PERFORMANCE OF SURFACE-PLASMA SOURCES

Type	Lab	Ref #	Arc I A	Arc V V	I(H-) mA	J(H-) A/cm2	e-/H- ratio	Pulse Hz, ms	Src P mT	Avrage H flow sccm	Emittance (fraction) pi-mm-mrad	Brightness (RMS,norm.) A/(m-rad)^2	Cesium Loss mg/hr	Source Volume cm^3	Aperture Diameter mm
Magnetron	BNL	6	10	160	65	1.1	.5-1	5,0.6	<100	1	1.1 (90%)	2.3e11	<1	1	2.8
	TAC	22	20	147	28 10	2.1 1.3		10,0.1 "				4.1e11		1 1	1.3 "
	FNAL	7	35	145	50	.5		15,0.085			1.0x1.5 (90) (at 750 keV)	1.43e11		1	1x10 slit
	DESY	9			40			8,0.14			1.5x1.0 (90)	1.14e11			
	ANL	8	40	145	40	.5		30,0.07			0.9x1.4 (90)	1.36e11		1	1x10 slit
Multicusp Converter	KEK	12	25-30	125	15-20			20,0.25 "	0.8	3-5	1.8-2.2 (90) (at 750 keV) 1.5 (90)	1.88e10 5.72e10	<1		
	LASL	11	60	100	20	.025		120,0.8		2	0.8 (95)	2.38e11	30	7226	10
	LBL	13	100	100	1250	.010	.12	DC	1.5	157.9			100	72000	45x250 slot
	Karlsr	24	35	70	50	.038		DC	1-2.5	2	0.4 rms	6.33e10	50-100	3016	13
Penning	Rüth.	16			30			50,0.5		13	1.7x3.5 (90) (at 665 keV)	2.16e10			
	LASL	15	180	92	50	.94	3	5,2			.049x.05 rms	4.14e12		4.6	2.6
	LASL	15	120	100	82	1.67	2.7	5,1		5	.053x.056 rms	5.60e12		.15	2.5
	LASL	25			60	1.5		5,0.05		3.1	.07x.15 rms	1.16e12			1x4 slit

202 H⁻ Ion Sources

TABLE 2 - PERFORMANCE OF VOLUME SOURCES

Lab	Ref #	Cs used?	Arc I A.	Arc V V	I(H-) mA	J(H-) A/cm2	e-/H- ratio	Pulse Hz, ms	P mT	Avrage H flow sccm	Emittance (fraction) pi-mm-mrad	Brightness (RMS,norm.) A/(m-rad)^2	Source Volume cm^3	Aperture Diameter mm
Triumf	26	No	30	145	1.85	.0056		DC	7	15	.038 rms	2.6e11	8168	6.5
Triumf	27	No	27	127	9	.008		DC			.44(90)	2.0e11	1178	12
"	"	No	"	"	7	.0074		"			.34(90)	2.6e11	"	11
KEK	23	No	100-200	120-150	3	.0068	>100			15-20		5.1e10	1260	7.5
"	"	Yes	"	"	20	.045	8				1.0 (90)		"	"
"	"	Yes	"	"	12	.027	<5		10-15				"	"
BNL	17	No	250		30	.030		0.5,1.3			.07 rms	5.4e11	1900	11.3
"	"	No	"		13	.013	"	"			.096 rms	4.2e11	"	"
"	"	No	"		19	.019			20				"	"
LASL/LBL	28	No	500	400	10	.032		60,0.8	9	15	.08 rms	3.2e11	6912	6.3
LBL	29	No			100	.015		10 ms	50				6912	29
LBL	30	No	350	150	.72	.365	>100	5,1					353	0.5
"	"	Yes	300	150	7.85	1.0	20	0.5 ms					"	1
"	"	Yes	350	150	150	1.2							"	19 apertures
Grumman	31	No			1.3	.165	300-	3,1	30		.01 rms	2.6e12	353	1
"	"	No			2.8	.09	600	"	9		.014 rms	2.9e12	"	2
"	32	Yes	300		2.6	.330	100	"			.014 rms	2.7e12	"	1
LBL	33	No			18	.0115		.25,270.			.137 rms	1.9e11	785	14
"	"	Yes			70	.0455					.35x.53(63)	3.1e11	"	"
Grumman	34	No	260	155	12	.037	97	3,1	25		.073 rms	4.6e11	785	6.4
"	35	No	260	155	2.9	.036	4.9	"	63		.04 rms	3.7e11	"	3.2
"	35	Yes		100	11.7	.145	10	"	30		.043 rms	1.3e12	"	3.2
"	36	Yes		130	22	.068	10	"	30		.065 rms	1.1e12	"	6.4
"	"	Yes			33	.102			30		0.11 rms	5.5e11	"	"
LBL	18	No	25 kW rf power		6	.19	55	150,1	15				785	2
LBL	37	No	50 kW rf power		31	.135			17				785	5.4
Grumman	34	No	40 kW rf power		18	.056	60		15		.08 (rms)	5.7e11	785	6.4

TABLE 3 - Summary of Characteristics of Various Type H⁻ Sources

	MAGNETRON	PENNING	MULTICUSP-CONVERTER	VOLUME	VOLUME + Cs
CURRENT DENSITY	High >1A/cm²	High >1A/cm²	Low 10-50 mA/cm²	Low-Medium 10-100 mA/cm²	Medium-High 50-1000 mA/cm²
BRIGHTNESS	Medium	High	Medium-Low	Medium	Medium
e / H RATIO	0.5-1	~ 2	< 0.5	5-500	1-50
SOURCE PRESSURE	≤ 100 mT	> 100 mT	1-3 mT	2-30 mT	
Cs CONSUMPTION	< 1 mg/hr.	≥ 1 mg/hr.	1-30 mg/hr.	0	?
DUTY FACTOR	Low	Low-Medium	High (cw)	High (cw)	High (cw)
DISCHARGE VOLUME	1.0 cm³	0.16 cm³ (SAS) 4.6 cm³ (4X)	~7000 cm³	25, 350, 800, 1200, 6900, 8200, 10600.... cm³	
OPERATIONAL EXPERIENCE ON ACCELERATORS	Most (BNL, FNAL, ANL, DESY)	Third (Rutherford)	Second (KEK, LAMPF)	Very Limited	

VIII. REFERENCES

1. Proceedings of the Fourth International Symposium on the Prod. and Neut. of Negative Ions and Beams, AIP Conf. Proc. 158 (1986).
2. Proceedings of the Fifth International Symposium on the Prod. and Neut. of Negative Ions and Beams, AIP Conf. Proc. 210 (1989).
3. P. Allison, ref. 1, p.465 (1986).
4. J.D. Lawson, "The Physics of Charged Particle Beams", Oxford Press, 1977.
5. J. Alessi and Th. Sluyters, Proc. Second Inter. Symp. on the Production and Neutralization of Negative Ions and Beams, BNL 51304, 153 (1980).
6. J.G. Alessi, J.M. Brennan, A. Kponou, Rev. Sci. Instrum. 61, 625 (1990).
7. C.W. Schmidt and C.W. Curtis, ref. 1, p.425 (1986).
8. V. Stipp, A. DeWitt, and J. Madsen, IEEE Trans. Nucl. Sci. NS-30, 2743 (1983).
9. H.S. Chang, G.-G. Winter, H. Krause, N. Schirm, and I. Tessmann, DESY HERA 87-04, February, 1987.
10. Yu.I. Belchenko, V.I. Davydenko, G.E. Derevyankin, G.I. Dimov, V.G. Dudnikov, I.I. Morosov, G.V. Roslyakov, A.L. Schabalin, Rev. Sci. Instrum. 61, 378 (1990).
11. R.R. Stevens, Jr., R.L. York, J. McConnell, R. Kandarian, 1984 Linear Accel. Conf., Seeheim, Germany, GSI-84-11, 226 (1984).

12. Y. Mori, A. Takagi, K. Ikegami, S. Fukumoto, ref. 1, p.378 (1986).
13. J.W. Kwan, G.D. Ackerman, O.A. Anderson, C.F. Chan, W.S. Cooper, G.J. DeVries, A.F. Lietzke, L. Soroka, W.F. Steele, Rev. Sci. Instrum. 57, 831 (1986).
14. J.D. Sherman, H.V. Smith, Jr., C. Geisik, P. Allison, "H⁻ Temperature Measurements by a Slit Diagnostic Technique", 1991 Part. Accel. Conf., San Francisco, CA, May, 1991 (to be published).
15. H.V. Smith, Jr., J.D. Sherman, P. Allison, 1988 Linear Accel. Conf., Newport News, Va., CEBAF Report 89-001, 164 (1989).
16. H. Wroe and N.D. West, Proc. 1988 European Part. Accel. Conf., Rome, Italy, 1439 (1988).
17. J.G. Alessi and K. Prelec, 1990 Linear Accel. Conf., Albuquerque, NM, Los Alamos Report LA-12004-C, 761 (1990).
18. K.N. Leung, G.J. DeVries, W.F. DiVergilio, R.W. Hamm, C.A. Hauck, W.B. Kunkel, D.S. McDonald, M.D. Williams, Rev. Sci. Instrum. 62, 100 (1991).
19. K.N. Leung, C.A. Hauck, W.B. Kunkel, S.R. Walther, Rev. Sci. Instrum. 60, 531 (1989).
20. S.R. Walther, K.N. Leung, W.B. Kunkel, Appl. Phys. Lett. 54, 210 (1989).
21. H. Kojima, M. Hanada, T. Inoue, Y. Matsuda, Y. Ohara, Y. Okumura, H. Oohara, M. Seki, K. Watanabe, Proc. 13th Symp. on Ion Sources and Ion-Assisted Technology, Tokyo, June, 1990.
22. C.R. Meitzler, P. Datte, F.R. Huson, P. Tompkins, 1990 Linear Accel. Conf., Albuquerque, NM, Los Alamos Report LA-12004-C, 710 (1990).
23. Y. Mori, T. Okuyama, A. Takagi, D. Yuan, Nucl. Instrum. Meth. A301, 1 (1991).
24. B. Piosczyk and G. Dammertz, Rev. Sci. Instrum. 57, 840 (1986).
25. P.G. O'Shea, T.A. Butler, M.T. Lynch, K.F. McKenna, M.B. Pongratz, T.J. Zaugg, 1990 Linear Accel. Conf., Albuquerque, NM, Los Alamos Report LA-12004-C, 739 (1990).
26. R. Baartman, K.R. Kendall, M. McDonald, P.W. Schmor, D. Yuan, 1987 IEEE Part. Accel. Conf., IEEE 87CH2387-9, 283 (1987).
27. K. Jayamanna, M. McDonald, D.H. Yuan, P.W. Schmor, 1990 European Part. Accel. Conf., Nice, France, 647 (1990).
28. R.R. Stevens, Jr., R.L. York, K.N. Leung, K.W. Ehlers, ref. 1, p.271 (1986).
29. J.W. Kwan, G.D. Ackerman, O.A. Anderson, C.F. Chan, W.S. Cooper, G.J. deVries, K.N. Leung, A.F. Lietzke, W.F. Steele, Rev. Sci. Instrum. 61, 369 (1990).

30. K.N. Leung, C.A. Hauck, W.B. Kunkel, S.R. Walther, Rev. Sci. Instrum. $\underline{60}$, 531 (1989).
31. T.W. Debiak, L. Solensten, J.J. Sredniawski, Y.C. Ng, R. Heuer, Rev. Sci. Instrum. $\underline{61}$, 392 (1990).
32. J. Sredniawski, L. Solensten, T.W. Debiak, Y. Ng, K.N. Leung, C.A. Hauck, Proc. First NPB Technical Symposium, Montcrey, CA, July 17-21, 1989.
33. J.W. Kwan, G.D. Ackerman, O.A. Anderson, C.F. Chan, W.S. Cooper, G.J. deVries, W.B. Kunkel, K.N. Leung, P. Purgalis, W.F. Steele, R.P. Wells, Rev. Sci. Instrum. $\underline{62}$, 1521 (1991).
34. G. Gammel, T. Debiak, J. Sredniawski, 1991 Part. Accel. Conf., San Francisco, CA, May, 1991 (to be published).
35. T.W. Debiak, L. Solensten, J. Sredniawski, Y.C. Ng, F. Keuhne, G. Gammel, Proc. Second NPB Technical Symposium, San Diego, CA, May 20-24, 1990.
36. J. Sredniawski, T. Debiak, G. Gammel, Grumman Accel. Technology Development Group Tech. Note. 91-03 (1991).
37. K.N. Leung, W.F. DiVergilio, C.A. Hauck, W.B. Kunkel, D.S. McDonald, 1991 Part. Accel. Conf., San Francisco, CA, May, 1991 (to be published).

SATURNUS: THE UCLA COMPACT HIGH-BRIGHTNESS LINAC

C. Pellegrini

UCLA Department of Physics
405 Hilgard Avenue
Los Angeles, California 90024-1547

ABSTRACT

Saturnus is a compact S-band linac designed to produce a single bunch with an energy of 20 MeV, peak current larger than 200A, energy spread of 0.3%, emittance smaller than 5 mm mrad (normalized rms), and a pulse duration of 4 ps or shorter. This bunch can be followed at a variable distance by a second weak "witness" bunch, to be used to measure the field excited by the first bunch,. The linac electron injector is based on the Brookhaven photoinjector design. Solenoids are used to focus the beam from the gun through a mirror box and a Plane Wave Transformer accelerating structure. The linac will be followed by two beam lines, one of them capable of longitudinal bunch compression, leading to experimental stations were beam-laser and beam plasma interactions will be studied.

INTRODUCTION

The production and characterization of high brightness, small emittance electron and positron beams is important for the development of the next generation of high luminosity colliders, for synchrotron radiation and coherent radiation sources, and other applications of accelerators. Although much progress has been done during the last few years with a better understanding of the limitations of electron sources and damping rings, and the introduction of laser driven photocathodes, we have not yet reached the level of beam quality needed in these applications. Furthermore in the future we will need bunches with very short, subpicosecond length, which are also not presently available.

A 20 MeV linac, Saturnus, is under construction at UCLA, in collaboration between the Physics and the Electrical Engineering Departments[1], to address some of these problems and more generally to support an experimental and theoretical program on Accelerator Physics at the UCLA campus. The main goals of this program are:
1. Production of low emittance, high brightness electron beams;
2. Development of compact linacs and new accelerating structures in S-band, and at higher frequencies;
3. Study of wake fields, shunt impedance and dark current effects in high frequency structures for high gradient accelerators;

4. Study of beam-plasma interactions, like plasma wake-field acceleration;
5. Study of the interaction of relativistic electron beams and high frequency electromagnetic radiation, as in FELs or in laser acceleration;
6. Training of graduate students in accelerator physics.

In this paper we will describe the linac, and its expected beam characteristics.

THE LINAC

The layout of the accelerator and experimental areas is shown in fig. 1, and a more detailed view of the linac in fig.2. The basic components of the system are:
1. a 5 MeV S-band photocathode RF gun;
2. a 15 MeV PWT accelerating section;
3. a Nd-Yag laser;
4. the beam transport lines and diagnostic;
5. the experimental area.

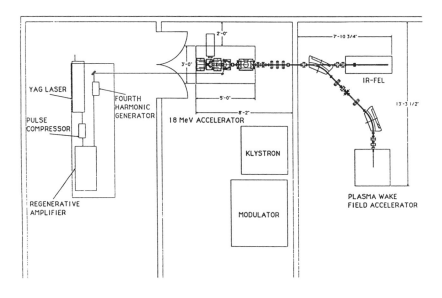

Fig. 1 Laboratory layout

The RF gun follows the Brookhaven design[2], with some modifications, like the possibility of injecting the laser beam at an angle of 70° or 2° to the axis of the photocathode, as shown in fig. 3. Its main parameters and the output beam characteristics are given in Table 1, and the design is shown in fig. 3.

TABLE 1	
Energy, MeV	4.5
Frequency, GHz	2.856
Number of cells	1 1/2
Maximum field on cathode, MV/m	100
Photocathode material	copper
Illumination	2^0 or 70^0
Emittance, mm mrad (Normalized, rms)	8
Pulse length, rms, ps	1.7
Charge, nC	1
Energy spread, %	0.3

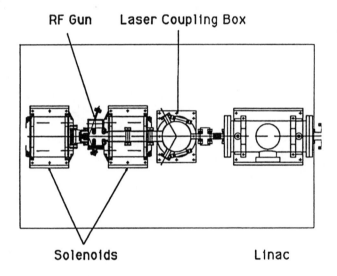

Fig:2 Accelerator components

The RF gun uses a metallic cathode. This has the disadvantage of a lower quantum efficiency than that obtainable from materials like cesium-antimonide used at Los Alamos[3], and the advantage of a longer lifetime and less restrictive vacuum requirements. The initial choice for the material is copper; however the cathode can be changed to test materials with larger quantum efficiency.

The laser system for the photoinjector will produce single or multiple pulses in the range of 2 to 30 picosecond with enough power to produce up to 10 nC per pulse from a copper photocathode. Its main elements are:

a Nd-Yag oscillator, mode locked at 38.08 MHz, with a pulse width smaller than 100 ps; a regenerative amplifier, with Silicate glass for large bandwidth, 5 Hz repetition rate and 50 mJ output per pulse; a pulse compressor using an optical fiber plus grating, producing a pulse width smaller than 4 ps; a fourth harmonic generator using KDP and BBO doubling crystals.

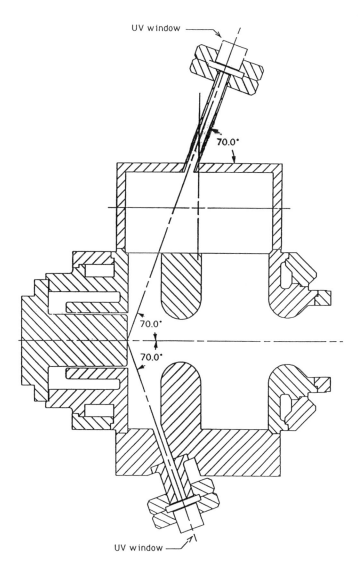

Fig. 3 Cross section of RF gun

The laser pulse can be injected either in a direction nearly perpendicular to the photocathode or at a 70 degrees angle. Illumination at this angle gives a larger quantum efficiency than illumination on axis; however the effective duration of the illumination is longer thus increasing the pulse length. In both cases we can simultaneously illuminate the cathode with the laser from two symmetric ports. The laser driving the photocathode will be able to provide single pulses or multiple pulses. In particular we intend to be able to produce a configuration with a large charge bunch followed by a witness bunch with smaller charge and emittance. The time separation between the two bunches is variables in steps corresponding to the RF period. We intend to use this laser for our initial work at the picosecond pulse level, and later to extend its performance to explore the production of subpicosecond electron pulses.

The gun is followed by a solenoid to compensate the beam angular divergence at the gun exit, and to focus the beam through the mirror box and an accelerating structure, the Plane Wave Transformer (PWT)[4], built in collaboration with D. Swenson. The solenoid is backed up by an equal magnet so that the field on the cathode is zero, to avoid an emittance increase.

Fig. 4 Cross section of PWT

The PWT is a standing wave structure with 8 cells; it consists of washers suspended inside a cylinder, as shown in a cross section of a 1/4 of a cell in fig. 4. The field is like that of a coaxial structure in the outer part, and is changed to a longitudinal E-field within the washers. This structure will accelerate the beam by 15 MeV in a length of about 40 cm; thus the full beam energy will reach 20 MeV. The PWT structure has some interesting properties, like easy mechanical construction and large shunt impedance; its wake field properties are also different from that of SLAC structures. We intend to study this structure using simulation codes, and by using it to accelerate a large charge bunch, followed by a small charge test bunch. This study is preliminary to similar studies on other structures, both in S-band and at larger frequencies, that we intend to pursue in the future.

The RF system is based on a RK5 klystron producing 25 MW peak power. A modulator with a 3.5 microsecond pulse length, the low power RF system, the electronics to synchronize the laser and the RF have been built at UCLA.

BEAM CHARACTERISTICS

The RF photoinjector, followed by the PWT structure will form a compact, about one meter long, linac producing a 20 MeV high brightness beam. The main beam characteristics, obtained from numerical simulations using Parmela, are given in Table 2 and shown in fig. 5.

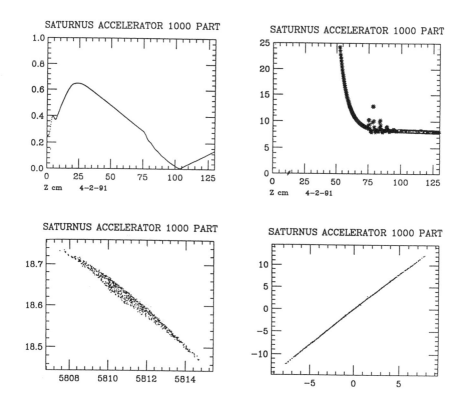

Fig. 5 Results of numerical simulations: 1. beam radius (rms) in cm, vs axial position, z, in cm, from the cathode to the PWT exit; 2. rms normalized emittance, in mm mrad, vs z; 3. beam energy, MeV, vs RF phase at linac exit; 4. horizontal angle, mrad, vs position, mm, at linac exit.

TABLE 2

Energy, MeV	20
Emittance, norm. rms, mm mrad	8
Peak current, A	200
Energy spread, rms	0.3
Bunch length, rms, ps	1.7
Repetition rate, Hz	5

As one can see from fig. 5 the beam characteristics predicted by the numerical simulations are quite good. In particular the bunch length remains essentially the same as the laser pulse length, the energy spread is a few times 0.1%, and the emittance is about 8 mm mrad. If the experimental results will be in agreement with the simulations, this beam will allow us to carry out an exciting program of particle beam physics.

Acknowledgements. I wish to thank the Brookhaven ATF group for their support and discussions during this work, and SLAC for providing us with the klystron and other RF parts. This work is being supported by DOE under grant DE-FG 03-90ER-40565.

REFERENCES

1. S. Hartman et al., Photocathode driven linac at UCLA for FEL and plasma wakefield experiments, Proc. 1991 Particle Accelerator Conference, San Francisco,
2. K. Batchelor et Al., Proc. 1989 Particle Accelerator Conf., Chicago, Il., p.273.
3. B. E. Carlsten and R. L. Sheffield, in Proc. Of 1988 Linear Accelerator Conference, Newport News, VA., p. 365.
4. D. A. Swenson, Proc. European Particle Accelerator Conf., Rome, Italy 1988, p.1418.

HIGH POWER MICROWAVE SOURCES FOR ADVANCED ACCELERATORS

V.L. Granatstein, P.E. Latham, W. Lawson, W. Main, M. Reiser,
C.D. Striffler and S. Tantawi
Laboratory for Plasma Research
University of Maryland, College Park, MD 20742

Abstract

The development of future linear colliders operating at $\gtrsim$ 1 TeV center of mass energy will likely depend on the availability of microwave amplifiers which are both affordable and are capable of significantly surpassing the performance of klystrons in terms of power and frequency. One of the most promising concepts for the required amplifiers is the gyroklystron. This paper reviews the physical principles of the gyroklystron and contrasts its phase bunching behavior with that of a klystron. Experiments in a low power gyroklystron which have studied phase noise and penultimate cavity tuning are summarized. The stages in developing an X-band gyroklystron resulting in the achievement of 29 MW, 1 μsec output pulses with 34% efficiency and 36 dB gain is recapitulated. Finally, plans are described for experiments which will extend gyroklystron performance to the $\sim$ 100 MW level at X-band, and explore operation at a frequency of $\sim$ 20 GHz by operating the output cavity at the second harmonic of the electron cyclotron frequency.

INTRODUCTION

The Stanford Linear Collider (SLC), which presently operates at 0.1 TeV center of mass energy, is driven by klystron amplifiers operating at a frequency of 2.85 GHz and producing 50 megawatt pulses with 3.5 μsec duration. It has been projected [1] that the next stage in linear collider development will aim at center of mass energy in the range 0.5 TeV to 1.5 TeV, and require microwave amplifiers at higher frequency and higher peak power than those used on SLC in order to keep the length of the accelerators within reasonable bounds. Microwave amplifiers required to drive this next stage linear collider would have frequency in the range 10-14 GHz and peak output power 50 - 150 megawatts with pulse duration $\sim 1 \mu\text{sec}$.[1] The pulses would be compressed by a factor of $\sim$ 8, and sources which trade-off power against pulse length might be acceptable.

In the more distant future, it is projected that for colliders with 3-5 TeV center-of-mass energy the microwave driver tubes will need to operate at frequencies in the range 20-30 GHz. The features of the SLC and projection for future colliders are summarized in Table I.

Extending the capability of klystron amplifiers to meet those advanced requirement presents severe and perhaps insolvable problems. A klystron amplifier is sketched in Fig. 1(a). The gaps in the cavities must be much smaller than the wavelength, and thus, scaling of this type of amplifier to higher power and shorter wavelength might be limited by rf breakdown in the output cavity gap.

An even more stringent limitation results from the fact that drift tube cross-section must be kept small enough to prevent the propagation of electromagnetic waves at the operating frequency. When this requirement is combined with a limited cathode emissivity, very large beam convergence is required in a high power, high frequency amplifier if voltage is to be kept $\lesssim$ 500 kV so that modulator cost does not become excessive. Large convergence angles increase the likelihood of electron interception on the drift tube walls.

Because of these klystron limitations alternate types of microwave amplifiers have been considered for application to advanced linear colliders. One of the most promising of these alternative amplifiers is the gyroklystron. Gyroklystrons are a specific embodiment of electron cyclotron masers which amplify coherent radiation near the electron cyclotron frequency, ω_{ce}. Their operation is based on the energy dependence of the cyclotron frequency; viz.

$$\omega_{ce} = \frac{eB_o}{m\gamma} \qquad (1)$$

where B_o is the amplified, dc axial magnetic field, and γ is electron energy normalized to mc^2. A gyroklystron is sketched in Fig. 1(b). Spiralling electrons produced by a magnetron injection gun stream though an input cavity where they gain or lose energy depending on their phase relative to the microwave signal in the cavity. The energy dependence of the cyclotron frequency as expressed in Eq. (1) then results in bunching of the electrons in phase in their cyclotron orbits as they pass through the drift space. The phase bunched beam entering the output cavity can efficiently transfer its energy to an electromagnetic wave whose resonant frequency in the output cavity is near ω_{ce}.

The nature of the phase bunching in a gyroklystron may be appreciated by considering Fig. 2. The cross-section of an annular beam of spiralling electrons is shown with beam radius r_b larger than the Larmor radius r_L. In Fig. 2(a), the situation in the input cavity of the gyroklystron is depicted with electrons 'smeared' in phase around their Larmor orbits. Also shown in Fig. 2(a) is an azimuthal electric field as would be present for a TE$_{01}$ mode excited in a cavity of circular cross-section. A the instant shown, an electron with phase like that of electron #2 will be accelerated by the electric field while electrons with phase like that of electron #1 will be decelerated. Overall there will be no net energy exchange in the input cavity but electrons will be energy modulated depending an their phase. Thus, in the drift space electron like #2 will have a higher value of γ and a lower value of ω_{ce} (see Eq. (1)), and accumulate relative phase lag. Whereas, electrons like #1 will accumulate phase lead. The result is that the electrons are phase bunched when they enter the output cavity as shown in Fig. 2(b). Then, effectively all the electrons can be decelerated by the electric field of a cavity wave at a frequency near ω_{ce}. Effective energy transfer from the electrons to the wave is thus accomplished.

Because electron bunching in the gyroklystron is in phase rather than in space no small gaps are required in the output cavity. Also, gyroklystrons are expected to have superior rejection of unwanted e.m. modes which do not resonate in the cavity near ω_{ce}. Thus, it would seem possible to develop gyroklystrons with relatively large cross-section whose cavities were resonant in a higher order mode, and whose drift space while cut off for the cavity mode would not necessarily have to be cut off for lower order modes. The suggestion that such a gyroklystron would scale to higher frequency and higher power than conventional klystrons was made in 1984,[2] and design studies were published in 1985.[3,4]

NOVEL PHYSICAL CHARACTERISTICS OF GYROKLYSTRONS

In the previous section, it was shown that the physical basis of the bunching mechanism in a gyroklystron is very different from that in a klystron (viz., bunching in phase of cyclotron orbits due to the relativistic energy dependence of ω_{ce} versus bunching in axial position). Because of this difference, it might be expected that the parametric dependence of important gyroklystron characteristics would be very different from that in conventional klystrons. One characteristic of special concern in the advanced accelerator application is phase noise.

The magnitude of phase fluctuation, $d\phi$, due to modulator voltage fluctuations of fractional magnitude dV/V has been analyzed[5] for both a conventional klystron and a gyroklystron. The equation showing the parametric dependence of the phase fluctuations in a klystron is[5]

$$\left(\frac{d\phi}{dV/V}\right)_{\text{klystron}} = \frac{\omega L}{c} \frac{1}{(\gamma+1)(\gamma^2-1)^{1/2}} \quad (2a)$$

The corresponding equation for phase noise in a gyroklystron is [5]

$$\left(\frac{d\phi}{dV/V}\right)_{\text{gyroklystron}} = \frac{\omega L}{c} \frac{(1+\alpha^2)^{1/2}}{(\gamma+1)(\gamma^2-1)^{1/2}} \left[\left(\frac{\gamma^2 \omega_{ce}}{\omega}-1\right) + \alpha^2 \gamma \left(1-\frac{\omega_{ce}}{\omega}\right)\right] \quad (2b)$$

In Eq. (2), L is the length of the amplifier circuit and α is the velocity ratio of the spiralling electrons in a gyroklystron (i.e. $\alpha = v_\perp/v_z$).

Equation (2a) for a klystron and (2b) for several values of α in a gyroklystron are plotted in Fig. 3. The dependence of phase fluctuation on beam voltage is seen to be radically different in the gyroklystron and the conventional klystron. In the conventional klystron, phase noise is very large in a low voltage device (viz., $d\phi/(dV/V) \to \infty$ when $V \to 0$); however, phase noise decreases monotonically with increasing beam voltage. On the other hand, for the gyroklystron, phase noise is very small in a low voltage amplifier (viz. $d\phi/(dV/V) \to 0$ when $V \to 0$); however, as beam voltage increases phase noises rises and asymptotically approaches a constant value which depends on α.

Also marked on Fig. 3 are experimental points for a 7.5 kV klystron,[5] a 30 kV gyroklystron,[5] and the 350 kV SLAC klystron.[6] The measurements are in good agreement with the theoretical curves. Thus, it may be expected that a high power gyroklystron operating with beam voltage in the range 400-500 kV and with $\alpha \approx 1$ would have phase noise greater than that in the 350 kV SLAC klystron by a factor ~ 3. It is judged that this level of phase noise can be handed,[7] especially because feedback techniques exist which effectively diminish phase noise.[8]

A second characteristic of the gyroklystron which contrasts sharply with the conventional klystron is the effect of detuning the penultimate cavity in a circuit with more than two cavities. In a conventional klystron, when the penultimate cavity is tuned to a resonant frequency somewhat higher than the signal frequency, the leading electrons at the head of a bunch are decelerated while the lagging electrons in the tail of the bunch are accelerated. The effect is to sharpen the bunch in space prior to the final drift tube and output cavity, and to enhance the gain and efficiency of the amplifier.

In a gyroklystron, however, bunch sharpening requires that the leading electrons in a bunch along the spiral orbit be accelerated so that γ increases, ω_{ce} decreases and the leading electrons slow down in their cyclotron motion. Similarly, the lagging electrons need to be decelerated. Thus, the sense of the required acceleration and deceleration to sharpen an electron bunch in a gyroklystron is opposite from what is required in a klystron. Therefore, in a gyroklystron, it was predicted[9] that one would need to detune the penultimate cavity to a resonant frequency below the signal frequency in order to enhance gain and efficiency.

In a recent experimental study,[10] using a 50 kW, 4.5 GHz, three-cavity gyroklystron, enhancing performance by detuning the penultimate cavity to a resonant frequency below the input signal frequency was demonstrated. In Fig. 4, output power, efficiency and gain of the amplifier are all plotted as a function of the penultimate cavity resonant frequency for a number of values of input power. All the curves are seen to have a broad maximum for detuning $\sim 0.6\%$ below the input signal frequency.

The gyroklystron on which this study was carried out with its 50 kW output power represented the state-of-the-art in stable gyroklystron operation until the 1990s. The development of a much more powerful gyroklystron whose operation at the ~ 30 MW level was demonstrated in 1991 is described in the next section.

EXPERIMENTAL DEVELOPMENT OF A 30 MW, 10 GHz GYROKLYSTRON

Beginning with the initial designs for a 30 MW, 10 GHz gyroklystron amplifier,[3,4] a systematic experimental development program was carried out at the University of Maryland aimed at realizing the design predictions. Initially, studies were carried out with a two-cavity circuit and much effort was expended in stabilizing spurious oscillations. This was accomplished primarily by judicious placement of lossy dielectric liners along the walls of the gyroklystron. Lossy dielectrics were introduced in the downtaper between the gun and the input cavity, in the input cavity itself, and in the drift tube between the cavities. The two cavity circuit was finally stabilized and optimized for maximum efficiency to achieve output power over 20 MW for 1 microsecond with peak power of 22 MW.[12,13] The applied modulator pulse flat top voltage was 425 kV and current was 150 A so that peak efficiency was 34%. Gain was 34 dB. To achieve this performance the solenoidal magnetic field was downtapered from a value of 0.54 T at the input cavity to 0.474 T at the output cavity. Figure 5 shows the variation in output power as the magnetic field taper is varied first with the input cavity field constant at 0.54 T and then with the output cavity field held constant at 0.474 T. With the magnetic field profile optimized for maximum output power, the value of α in the output cavity was near one.

Recently, a three cavity gyroklystron configuration has been used to increase gain to 50 dB with little degradation of output power or efficiency.[14] The three cavity gyroklystron configuration is sketched in Fig. 6. The penultimate cavity in this circuit was tunable and optimum performance was achieved by detuning to a resonant frequency below the signal frequency as described in the previous section. Gain and efficiency are plotted vs. penultimate cavity detuning in Fig. 7. The gain curve is seen to be sharply peaked and gain approaching 50 dB is shown. In Fig. 8, efficiency and output power are plotted vs. beam current for two different values of magnetic field downtaper. It is seen that with a 33% downtaper efficiency is almost 40% with a current of 100 A, while at 200 A output power approaches 30 MW with efficiency still above 30%. Generally, some feature of the amplifier performance could be optimized by adjusting the magnetic field downtaper, the penultimate cavity detuning, the beam current and the value of α. Table II displays the performance achieved with three different adjustments of the amplifier settings, one for maximum efficiency, one for maximum gain, and one for maximum power. The maximum gain setting with measured gain of 50 dB clearly demonstrates that gyroklystron gain can be enhanced effectively by adding a third cavity with little degradation in efficiency and power compared with the two cavity gyroklystron performance.

The maximum power adjustment is also interesting in that an especially wide pulse was achieved together with high peak power. The energy in the microwave output pulse (41 J) is considerably beyond what has been achieved[15] in experimental studies of conventional klystrons operating in X-band. Finally, we note that maximum power was achieved with a very large magnetic field downtaper of 33% resulting in a value of $\alpha = 0.71$ in the output cavity. Such a low value of α implies that a significant fraction of the microwave output power of 29 MW must come from the axial motion of the electrons. If all the energy had come from the transverse motion, transverse efficiency would have to have been >100% which is of course impossible. In most gyrotrons operating with beam voltage $\lesssim$100 kV energy is extracted primarily from the transverse electron motion. However, in the 425 kV gyroklystron the physical process of energy transfer from the beam to the microwaves is evidently different, and this is currently under study.

FUTURE PROSPECTS

The demonstration of stable gyroklystron operation at an output power level of 29 MW in $\gtrsim 1\mu$sec pulses must be considered auspicious, especially since the power level was near the design value.[3,4] It also represented more than a five hundred fold enhancement of power level over that in previous, thermionic cathode, gyroklystron studies.[16,17]

Average power capability of gyroklystrons has not yet been addressed. Since the collider application will require a pulse repetition frequency of several hundred hertz, average output power rating would need to be about 0.03% of the peak power rating. Water cooling of the copper walls of the output cavity and beam collector would be well within the state of the art. Cooling of the lossy dielectric in the penultimate cavity is a new problem and must be addressed; however, the lower field amplitudes here compared with the output cavity should make the problem manageable.

A new gyroklystron experiment which is presently being designed involves modifying the output cavity in the amplifier described in Section III so that it will be excited in the TE°_{021} mode at the second harmonic of the cyclotron frequency. Output power is predicted to be greater than 10 MW at a frequency near 20 GHz.

Given that this 20 GHz experiment agrees with predictions, and verifies the possibility of efficient, stable, second harmonic operation, a new gyroklystron will be developed at a frequency of 11.4 GHz and power in the range 100–150 MW with 1 μsec pulses; these performance characteristics correspond to the perceived requirements in Table I. The input cavity and penultimate cavity will be designed to operate at the fundamental frequency (i.e. 5.7 GHz) so that a larger circuit cross-section will be used compared with the X-band circuit. Beam voltage will be increased by only to 500 kV. The circuit will be coaxial with the center conductor functioning to diminish the density of spurious modes; also if the center conductor is clad with lossy dielectrics it can further improve stability.

At this point, one can be guardedly optimistic that the rf source requirements of future linear colliders will be met in a timely manner. The gyroklystron appears to be a leading approach, but other promising concepts are also being explored ranging from extending the operating range of the well known-TWT amplifier[18] to new configurations involving spiralling pencil beams.[19] The pace of collider development may influence the choice of rf sources in addition to the scientific and technical progress with the sources themselves.

ACKNOWLEDGEMENTS

This work was supported by the U.S. Department of Energy, Division of High Energy Physics.

REFERENCES

1. R. Ruth, ed., Report of the Linear Collider Working Group, Proceedings of the 1990 Summer Study on High Energy Physics, Snomass, CO, June 25-July 13, 1990.

2. V.L. Granatstein, Int. J. Electronics 57, 787 (1984).

3. V.L. Granatstein, P. Vitello, K.R. Chu, K. Ko, P.E. Latham, W. Lawson, C.D. Striffler and A. Drobot, IEEE Trans. Nucl. Sci. NS-32, 205 (1985).

4. K.R. Chu, V.L. Granatstein, P.E. Latham, W. Lawson and C.D. Striffler, IEEE Trans. Plasma Sci. PS-13, 424 (1985).

5. G.S. Park, V.L. Granatstein, P.E. Latham, C.M. Armstrong, A.K. Ganguly and S.Y. Park, IEEE Trans. Plasma Sci. 19, 632 (1991).

6. T.G. Lee, private communication.

7. P.B. Wilson, private communication.

8. G.S. Park, V.L. Granatstein, J. McAdoo, C.M. Armstrong, A.H. McCurdy and S.Y. Park, "Phase Noise Reduction in a Gyroklystron Amplifier," Int. J. Electronics (to be published).

9. K.R. Chu, P.E. Latham, and V.L. Granatstein, Int. J. Electronics 65, 419 (1988).

10. G.S. Park, V.L. Granatstein, S.Y. Park, C.M. Armstrong, and A.K. Ganguly, "Experimental Study of Efficiency Optimization in a Three-Cavity Gyroklystron Amplifier," submitted to IEEE Trans. Plasma Sci.

11. J.P. Calame, W. Lawson, V.L. Granatstein, P.E. Latham, B. Hogan, C.D. Striffler, M.E. Read, M. Reiser, and W. Main, J. Appl. Phys. 70, 2423 (1991).

12. W. Lawson, J.P. Calame, B. Hogan, P.E. Latham, M.E. Read, V.L. Granatstein, M. Reiser, and C.D. Striffler, Phys Rev. Lett 67, 520 (1991).

13. W. Lawson, J.P. Calame, B. Hogan, M. Skopec, C.D. Striffler, V.L. Granatstein, and W. Main, "Performance Characteristics of a High Power, X-Band, Two-Cavity Gyroklystron," submitted to IEEE Trans. Plasma Sci.

14. S. Tantawi, W. Main, P.E. Latham, G. Nusinovich, B. Hogan, H. Matthews, M. Rimlinger, W. Lawson, C.D. Striffler, and V.L. Granatstein, "High Power X-Band Amplification from an Over-Moded, Three-Cavity Gyroklystron with a Tunable Penultimate Cavity," submitted to IEEE Trans. Plasma Sci.

15. A. Vlieks, R.S. Callin, C. Caryotakis, K.S. Font, W.R. Fowkes, T.G. Lee, and E.L. Wright, SLAC-PUB-5480 (also, to be published in Proc. 1991 IEEE Particle Accelerator Conference).

16. R.S. Symons and H.R. Jory, Advances in Electronics and Electron Physics, Vol. 55, L. Marton and C. Marton, eds. (Academic Press, New York, 1981) pp. 1-75.

17. W.M. Bollen, A.H. McCurdy, B. Arfin, R.K. Parker, and A.K. Ganguly, Trans. IEEE Plasma Sci. PS-13, 417 (1985).

18. D. Shiffler, J.A. Nation, L. Schachter, J.D. Ivers, and G.S. Kerslick, J. Appl. Phys. 70, 108 (1991).

19. O. Nezhevenko, "The Magnicon: a new rf power source for accelerators," Proc. 1991 IEEE Particle Accelerator Conference (in press).

Table I. RF source requirements for some projected linear colliders[1]
(The SLC is presently operating and is included for comparison.)

System	CM Energy (TeV)	f_{rf} (GHz)	P_{rf} (MW)	τ_{rf} (μs)
SLC	0.1	2.85	50	3.5
NLC1	0.5	11.424	50	0.9*
NLC2	1.0	11.424	100	0.9*
NLC3	1.5	11.424	150	0.9*
Future	≥3.0	20-30		

*Pulses would be compressed 8× before coupling into accelerator.

Table II. Parameters of the 3-cavity gyroklystron when adjusted to optimize some output feature.

	Max. Efficiency	Max. Gain	Max. Power
Efficiency	39%	31%	34%
Output power	17 MW	21 MW	29 MW
Microwave pulse energy	24 J	22 J	41 J
Gain	33 dB	50 dB	36 dB
Beam current	100 A	166 A	204 A
Penultimate detuning	−66.4 MHz	−42.2 MHz	−78.8 MHz
Magnetic field taper (reduction from input to output cavity)	30%	17%	30%
B_z (output cavity)	4.58 kG	4.53 kG	4.58 kG
α (output cavity)	0.84	0.73	0.71

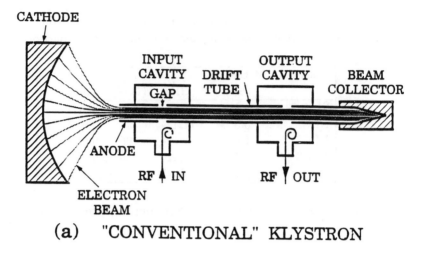

(a) "CONVENTIONAL" KLYSTRON

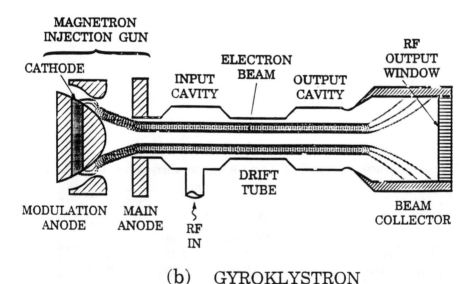

(b) GYROKLYSTRON

Figure 1: (a) Sketch of a 2-cavity klystron amplifier showing the strong beam convergence from the cathode to the circuit required in a high power, high frequency amplifier. (b) Sketch of a 2 cavity gyroklystron amplifier. The beam convergence required is much less than for the klystron as shown.

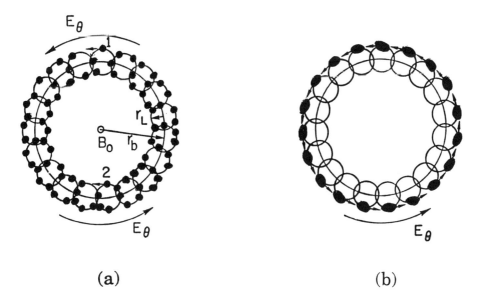

Figure 2: Phase bunching in a gyroklystron operating the TE_{01} mode in cavities with circular cross-section. (a) Situation in the input cavity with electrons 'smeared' in phase in their cyclotron orbits. (b) Situation in the output cavity showing electrons bunched in phase.

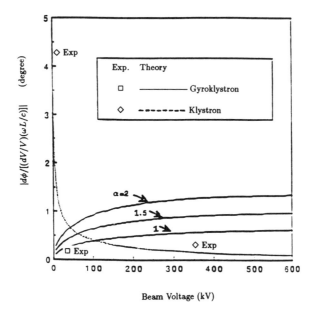

Figure 3: Phase fluctuations of a klystron and gyroklystron vs. beam voltage. (Adapted from Fig. 5 in reference 5.)

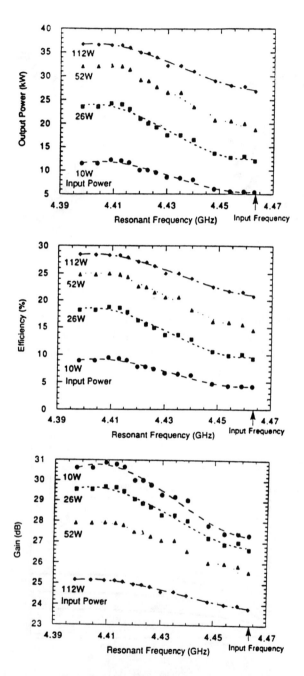

Figure 4: Output power, efficiency and gain of a three-cavity gyroklystron as a function of the penultimate cavity resonant frequency. (Fig. 7 in reference 10.)

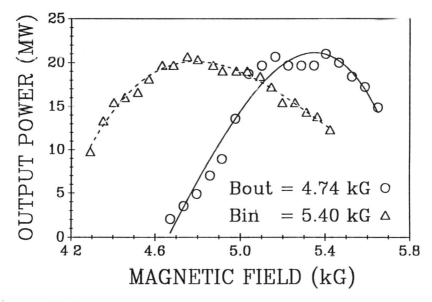

Figure 5: Peak output power vs. magnetic field variation in the input or output cavities as the magnetic field taper is changed in the two-cavity gyroklystron (Fig. 7 in reference 12).

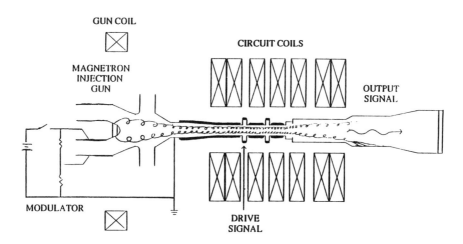

Figure 6: Schematic diagram of the 3-cavity gyroklystron. Darkened regions along the tube walls indicate the position of lossy dielectric liners. (Adapted from Fig. 1 in reference 14.) All cavities operate in the TE_{011}^o mode.

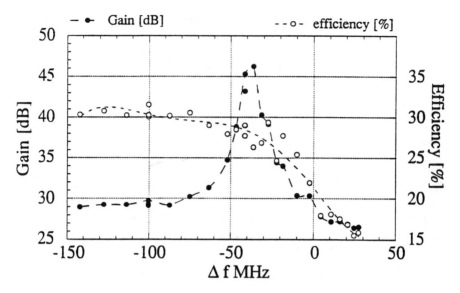

Figure 7: Effect of penultimate cavity detuning on the performance of the 3-cavity gyroklystron (magnetic field was downtapered by 17% from the input cavity to the output cavity, beam voltage was 425 kV, beam current was 160 A (Fig. 20 in reference 14.)

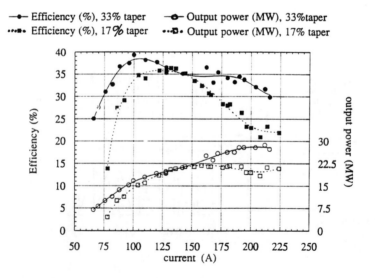

Figure 8: Output power and efficiency vs. beam current for magnetic field tapers of 33% and 17% (Fig. 18 in reference 14.)

PANEL DISCUSSION
M. Reiser, Chairman

The Panel, whose members included (in addition to the Chairman) Don Edwards, Robert Gluckstern, Terry Godlove, Victor Granatstein, Robert Jameson, Pierre Lapostolle, John Lawson, Ronald Martin, and Claudio Pellegrini, provided a forum to highlight, discuss and clarify some of the issues presented in the talks and to contribute additional information relevant to the Symposium.

The Chairman started the Panel Discussion with a comparison of beam requirements for various future advanced accelerators discussed in the Symposium and the current state-of-the-art and with a brief review of the physics research issues relevant to the generation of high-brightness beams. A written summary is given below. John Lawson discussed some additional aspects of the emittance concept that he included in the written version of his talk. Pierre Lapostolle made a few remarks on halo formation and equipartitioning (see below). Robert Gluckstern contributed some observations on beam mismatch and longitudinal - transverse coupling in a linac (see below). Ronald Martin gave a valuable personal account of the history of non-Liouvillean injection and its connection to heavy ion fusion, a written version of which is presented below.

One of the other topics discussed by the Panel was the different definitions of emittance: $\epsilon = 6\epsilon_{rms}$ at Fermilab (Gelfand), $\epsilon = \epsilon_{rms}$ at the SSC Laboratory (Edwards), and $\epsilon = 4\epsilon_{rms}$ (first proposed by Lapostolle in 1971) at many other places. Clearly, a consensus on this matter would be highly desirable, and perhaps a future symposium could address this issue.

On the question of high-brightness beam requirements for short-wavelength free electron lasers, Claudio Pellegrini pointed out that the effective overall beam emittance is not necessarily the important parameter. Rather, what counts most is that the high-density core of the beam overlaps with the laser beam and thereby provides the desired gain.

Following below are the written accounts of the contributions to the Panel Discussion by the Chairman, Pierre Lapostolle, Robert Gluckstern, and Ronald Martin.

M. REISER:

I have prepared two tables in which I have tried to highlight the major topics discussed at this Symposium. In the first (Table I) I listed the *Advanced Accelerator* projects and concepts that we heard about and compared the beam requirements ("future challenge") with the "current state of the art". The numbers are to be taken as ballpark figures and may not always represent the latest results or design parameters. Nevertheless, they give an indication of the challenges that are ahead in achieving the desired performance goals (energy, emittance, brightness, luminosity, power, etc.) for the various advanced accelerator projects. The names of the Symposium speakers who talked about the relevant topic or accelerator are indicated.

The second table lists the major *Beam Physics Research Issues* relevant to the generation of high-brightness beams. Those discussed at the Symposium are indicated by the names of the speakers. There were no talks on the conventional cooling methods (stochastic, electron beam, radiation) employed in high-energy colliders to reduce the emittance. However, some new ideas on this topic were presented by Phil Sprangle (laser cooling) and Yaroslav Derbenev (self cooling). Also, the very important field of instabilities was not covered at this symposium. Both topics, cooling and instabilities, should be included in a future symposium of this type.

I. ADVANCED ACCELERATORS

The Future Challenge	The Current State of the Art
SSC 20 TeV, $\mathcal{L} = 10^{33}$ cm^{-2}s^{-1} Normalized rms emittance $\tilde{\epsilon}_N \approx 1$ mm-mrad 3 boosters D. Edwards	**Tevatron** 0.9 TeV, $\mathcal{L} \sim 10^{30}$ cm^{-2} s^{-1} $\tilde{\epsilon}_N \approx 1.5$ mm-mrad 1 booster N. Gelfand
NLC 250 (750) GeV × 250 (750) GeV $\mathcal{L} = 2 \times 10^{33}$ cm^{-2}s^{-1} $\tilde{\epsilon}_N \lesssim 0.1$ mm-mrad $\sigma_y \sim 20$Å, $\sigma_x \sim 100\sigma_y$ J. Seeman	**SLC** 50 GeV × 50 GeV $\mathcal{L} \leq 10^{29}$ cm^{-2}s^{-1} $\tilde{\epsilon}_N \sim 50$ mm-mrad $\sigma_x = \sigma_y \gtrsim 1\mu$m
HIF Drivers Beam requirements at target: 5-10 GeV, 20 kA, 10 ns $\tilde{\epsilon}_N \sim 10$ mm-mrad I. Hofmann, T. Fessenden	**Experiments** ~500 keV, 0.1-1 A pulse length $\sim \mu$s $\tilde{\epsilon}_N \sim .05$ mm-mrad I. Hofmann, T. Fessenden
Radioactive Waste Transmutation 1.6 GeV (p) 250 mA average current 400 MW average power $\tilde{\epsilon}_N \sim 1$ mm-mrad R. Jameson	**LAMPF** 0.8 GeV (p) 1 mA average/17 mA peak $\lesssim 1$ MW average power $\tilde{\epsilon}_N \gtrsim 1$ mm-mrad
FEL $\lambda \lesssim 1$ mm High Power (MW) 0.1 - 1 kA $\tilde{\epsilon}_N \sim \lambda$ ($\ll 1$ mm-mrad) C. Roberson, C. Pellegrini	**LLNL Experiment** $\lambda \sim 10$ mm $\hat{P} \sim 200$ MW 3 MeV, 1 kA $\tilde{\epsilon}_N \gtrsim 50$ mm-mrad
High Power Microwave Sources Need $\gtrsim 10$ GHz, $\gtrsim 300$ MW $\gtrsim 100$ ns, $\gtrsim 100$ Hz for linear colliders V.L. Granatstein	**Experiments** ~ 10 Ghz ~ 200 MW, 50 ns – single pulse (LLNL) ~ 25 MW, 1 μs (SLAC, U of MD)

II. BEAM PHYSICS RESEARCH ISSUES

Sources (J. Alessi, P. O'Shea)

Increase current densities to achieve lower emittance, higher brightness, e.g. photocathodes ($J \sim 600$ A/cm^3, 30 times J of thermionic cathodes), H$^-$ sources, heavy ions

Emittance Growth (P. Lapostolle, T. Wangler, O. Anderson, D. Kehne et al.)

Space-charge induced emittance growth, halo formation, mismatch at transitions
<u>Instabilities</u>: not discussed at Symposium (except J.G. Wang et al.)
New look at emittance concept (<u>J.D. Lawson</u>)

Non-Liouvillean Stacking (I. Hofmann, GSI)

How much can we gain before reaching space-charge limits (tune shift, instabilities)?
HIF Driver concept

Cooling (Not discussed at Symposium)

Electron-beam, stochastic for $p, \bar{p}$
Radiation for e^+, e^-
Is it possible to cool in <u>short</u> distance (time) rather than using expensive rings?
(P. Sprangle: Laser cooling; Ya. S. Derbenev: Self-cooling of beams)

Nonlinear Dynamics

Solitons – an intriguing possibility (J. Bisognano)
Nonlinear effects in beam transport:
        Correction of aberrations (A. Dragt)
        Emittance growth in ESQ transport? (S. Guharay et al.)
Wide-open field of research!

P. LAPOSTOLLE:

Amongst space charge effects which have not yet been completely mastered there are some which may remain a technical problem for the future. One of them is the halo formation; this one will become a serious concern with its associated risk of losses for the high average intensity machines as presented by R. Jameson for the radioactive waste transmutation project.
Several causes for this halo formation are, for instance:
a mismatch;
a bad distribution of the particles;
a misalignment (without and with aberrations and image effects).
Some of these effects are well known and mastered; other ones might still require some additional study.
One may also mention equipartitioning; this is perhaps a mechanism contributing also to the halo development. Its existence has been clearly demonstrated and some of its deleterious effects mastered; it is not sure, however, that a better understanding of its process could not give another insight into other phenomena; some new guidelines for the reduction of the halo and its resulting risk of beam loss could even appear.

R.L. GLUCKSTERN:

I have just a couple of observations to make:

1. Consider the envelope equation in 1-D

$$a'' + k_t^2 a = \frac{\epsilon_t^2}{a^3} + \frac{K}{a},$$

where

$$K = \frac{2e\tau}{4\pi\epsilon_0 m v^2}.$$

In the large space charge limit, the emittance term can be neglected and the matched radius is given by

$$a_0^2 = \frac{K}{k_t^2}.$$

The frequency of small oscillation about the matched radius is

$$k_a^2 = k_t^2 - \frac{K}{a_0^2} = k_t^2 - \frac{\omega_p^2}{2v^2},$$

where $n = \tau/\pi a_0^2 e$ is the number of charges per unit volume and where $\omega_p^2 = ne^2/m\epsilon_0$ is the square of the plasma frequency. In fact, all beam oscillation modes will have frequencies of order of the plasma frequency. Thus a "mismatch" will change the charge distribution on a time scale of order of ω_p^{-1}, as observed in the simulations.

2. The second point is illustrated by examining the effect of the longitudinal transverse coupling in a linac, arising from the $(x^2+y^2)z$ term in the Hamiltonian. If the transport channel is matched for the synchronous particle, a nonsynchronous particle will lead to an elliptical distortion of the "matched circle" in the transverse phase space, where the magnitude of the elliptical distortion is proportional to the magnitude of the initial longitudinal oscillation, and where its orientation depends on the initial phase of the longitudinal oscillation. A projection will therefore occupy a greater phase space area, even though the original 6-D volume is not changed. This suggests that it may ultimately be possible to reduce emittance growth in the projections by introducing the appropriate initial 6-D phase space correlations. Of course this may turn out to be very difficult to implement, but it is one further indication that some emittance growth is not inevitable.

Finally, I wish to add that my interactions with Martin Reiser here at UMCP and away from home at Los Alamos, where we have spent time together, have been a pleasure. Happy birthday, Martin.

R.L. MARTIN: *History of Non-Liouvillian Injection and Its Connection to the Initiation of the U.S. Program in Heavy Ion Fusion*

The first suggestion that I am aware of the possibility of non-Liouvillean injection was by Moon from Birmingham in 1956. It was published in the Proceedings of the First International Accelerator Conference held in CERN that year. He suggested the molecular dissociation of H_2^+ to give protons. I am not aware that it was ever tried experimentally.

I visited Novosibirsk in late 1968 and saw a laboratory experiment by G. Dimov on charge exchange injection of H^- ions. He injected 1.5 Mev H^- ions into a small storage ring (~ 2 m in diameter) in a two stage process ($H^- \to H^0 \to H^+$) by gas stripping. Although he only accumulated a few $\times 10^{10}$ protons this was impressive because it was at the space charge limit of the ring at 1.5 MeV and required many turns of injection. It was therefore clearly non-Liouvillean. Even more impressive to me was that Dimov had developed an H^- source of 16 mA output. (Dudnikov, for whom a source type is named today, was part of Dimov's group.) This was far superior to any other H^- source in the world at that time, and the current was adequate to make charge exchange injection practical on an operating physics machine at high intensity.

I therefore began a program on H^- charge exchange at Argonne where the ZGS injector was a 50 MeV Alvarez Linac. At 50 MeV one could make plastic foils thin enough (2000 Å) and that the complication of gas stripping could be avoided. Major questions were:
1) would lifetime of foils be long enough to make the operation acceptable
2) how much gain in brightness was achievable
3) could one reach high intensities (5 - 10 $\times 10^{12}$ protons/pulse).

The answer was subsequently shown to be positive on all these issues.

Argonne's first test was injection of H⁻ into the 12 GeV ZGS in 1970. The injected current at that time was very low (1/4 mA) so the current was not adequate for physics operation of the ZGS. Nevertheless, the injection efficiency was remarkable, foils worked, and it spurred on the program.

This led to acquisition of the 2.2. GeV Cornell electron synchrotron to use as a test bed when its operation at Cornell was terminated. Operating at a proton energy of 200 MeV, we were able to demonstrate that H⁻ charge exchange injection was a practical and advantageous injection method (as well as carrying out on this machine prototype work on both intense pulsed neutron sources and medical applications of protons). Brightness increases of the circulating beam over that of the injected beam of a factor of 100 were achieved. Construction of a rapid-cycling (30 Hz) 500 MeV proton synchrotron was begun for the purposes of injection into the ZGS and as the source of protons for an intense pulse neutron source (IPNS). Operation of the ZGS was terminated before the booster could be used as an injector, but the booster is currently in operation for IPNS.

Charge exchange injection into the ZGS was carried out from 1975-76, and into the booster in 1977. It was primarily responsible for doubling the ZGS intensity to 6×10^{12} protons/pulse and was used operationally until the ZGS was turned off for non-polarized proton physics. While the brightness increase was expected, a most pleasant surprise was the enhanced stability of the synchrotron. All pulses were very near the peak intensity with very little knob twiddling, including the first accelerated pulse after a breakdown.

Charge exchange injection into the booster, now referred to as the Rapid-Cycling Synchrotron (RCS), made the RCS the most intense proton synchrotron in the world until it was exceeded a few years ago by the ISIS synchrotron in Rutherford.

Charge exchange injection was subsequently utilized on many synchrotrons around the world. These include the boosters at KEK and Fermilab, the AGS at Brookhaven, the Proton Storage Ring (PSR) at Los Alamos, and the ISIS synchrotron at Rutherford. All use foil stripping of H⁻ except the PSR, which utilizes magnetic stripping.

This experience with H⁻ charge exchange injection was what stimulated my ideas on inertial confinement fusion, first with protons, then α particles, and finally iodine, all of which involve non-Liouvillean injection. Since these concepts were developed in parallel with ideas on heavy ions for inertial confinement fusion by Al Maschke some detail in this evolution might be in order. The time table that I recall is enclosed.

February 1974	Maschke internal BNL report on high dE/dx of high energy U in DT–report classified.
April 1974	Martin–Protons better than electrons for ICF, and with charge exchange injection could get 30,000 A at 50 MeV in 100 small rings.
July/August 1974	Martin discussed at Aspen, Protons adequate?

March 1975	PAC Meeting in Washington
	    Martin – Charge exchange of He^+.
	    Maschke – dE/dz of U in DT.

April 1975	Martin talk about ICF at Sandia. Clauser: not adequate.
Summer 1975	Sea Ranch (California) Meeting Maschke – 100 GeV storage ring with U can contain 100 MJ for ICF. Martin – Know how to solve injection problem with alphas, not with heavier ions.
Fall 1975	Martin, Arnold – Molecular dissociation of hydrogen iodide with doppler shifted ruby laser. $\sim$1 MJ of I at 40 GeV with 98 beams from 1 ring by stripping extraction.
February 1976	Meeting in Germantown (or Rockville) to discuss hydrogen oxide concept and ICF in general. DOE decides to organize workshops.
March/April 1976	Keefe on Induction Linacs for purpose.
July 1976	First Heavy Ion Fusion workshop at Berkeley a) Energies too high. b) Intense sources of singly charged ions available, don't need non-Liouvillean injection. c) Interesting – DOE will fund.

NON-LIOUVILLEAN INJECTION

$H_2^+ \overset{\overline{\rightarrow}}{\text{gas}} H^+$ Moon (Birmingham, 1956)

$H^- \overset{\rightarrow}{\text{gas}} H^+$ Dimov (Novisibirsk, 1968)

$H^- \overset{\rightarrow}{\text{foil}} H^+$ Martin (ANL, 1970-76)
| 100 × Brightness | ZGS
| Stability | ZGS BST (IPNS)
KEK BST
FNAL BST
AGS (BNL)
PSR (LANL) (field stripping)
ISIS (Rutherford)

$HI^+ \overset{\rightarrow}{\text{laser}} I^+$ Martin, Arnold (1976) HIF

$Bi^+ \longrightarrow Bi^{2+}$ Rubbia (1989) HIF Main Ring

$Ba^+ \longrightarrow Ba^{2+}$ Hofmann (1990) HIF Compressor Rings

— — — — — — — — — — — — — — — — — — — —

Extraction $H^- \overset{\rightarrow}{\text{foil}} H^+$ TRIUMPF (Vancouver)

Proposed for ACCTEK Proton Theraphy,
Proton Radiography and Computed Tomography

AUTHOR INDEX

A
Alessi, J. G., 193
Allen, C. K., 67
Anderson, O. A., 66

B
Bisognano, J. J., 42

D
Derbenev, Y. S., 103

E
Edwards, D. A., 122
Esarey, E., 87

F
Fessenden, T. J., 160

G
Gelfand, N. M., 111
Granatstein, V. L., 213
Guharay, S. K., 67
Guo, W. M., 57

H
Hafizi, B., 170
Hofmann, I., 149

J
Jameson, R. A., 139

K
Kehne, D., 47
Kwan, J. W., 66

L
Lapostolle, P., 11
Latham, P. E., 213
Lawson, J. D., 1
Lawson, W., 213

M
Main, W., 213

O
O'Shea, P. G., 182

P
Pellegrini, C., 206

R
Reiser, M., 47, 57, 67, 213, 227
Roberson, C. W., 170
Rudd, H., 47

S
Seeman, J. T., 129
Soroka, L., 66
Sprangle, P., 87
Striffler, C. D., 213
Syphers, M. J., 122

T
Tantawa, S., 213

W
Wang, D. X., 57
Wang, J. G., 57
Wangler, T. P., 21
Weis, T., 77

AIP Conference Proceedings

		L.C. Number	ISBN
No. 101	Positron-Electron Pairs in Astrophysics (Goddard Space Flight Center, 1983)	83-71926	0-88318-200-9
No. 102	Intense Medium Energy Sources of Strangeness (UC-Santa Cruz, CA, 1983)	83-72261	0-88318-201-7
No. 103	Quantum Fluids and Solids – 1983 (Sanibel Island, FL)	83-72440	0-88318-202-5
No. 104	Physics, Technology and the Nuclear Arms Race (APS, Baltimore, MD, 1983)	83-72533	0-88318-203-3
No. 105	Physics of High Energy Particle Accelerators (SLAC Summer School, 1982)	83-72986	0-88318-304-8
No. 106	Predictability of Fluid Motions (La Jolla Institute, 1983)	83-73641	0-88318-305-6
No. 107	Physics and Chemistry of Porous Media (Schlumberger-Doll Research, 1983)	83-73640	0-88318-306-4
No. 108	The Time Projection Chamber (TRIUMF, Vancouver, 1983)	83-83445	0-88318-307-2
No. 109	Random Walks and Their Applications in the Physical and Biological Sciences (NBS/La Jolla Institute, 1982)	84-70208	0-88318-308-0
No. 110	Hadron Substructure in Nuclear Physics (Indiana University, 1983)	84-70165	0-88318-309-9
No. 111	Production and Neutralization of Negative Ions and Beams (3rd Int'l Symposium) (Brookhaven, NY, 1983)	84-70379	0-88318-310-2
No. 112	Particles and Fields – 1983 (APS/DPF, Blacksburg, VA)	84-70378	0-88318-311-0
No. 113	Experimental Meson Spectroscopy – 1983 (7th International Conference, Brookhaven, NY)	84-70910	0-88318-312-9
No. 114	Low Energy Tests of Conservation Laws in Particle Physics (Blacksburg, VA, 1983)	84-71157	0-88318-313-7
No. 115	High Energy Transients in Astrophysics (Santa Cruz, CA, 1983)	84-71205	0-88318-314-5
No. 116	Problems in Unification and Supergravity (La Jolla Institute, 1983)	84-71246	0-88318-315-3
No. 117	Polarized Proton Ion Sources (TRIUMF, Vancouver, 1983)	84-71235	0-88318-316-1
No. 118	Free Electron Generation of Extreme Ultraviolet Coherent Radiation (Brookhaven/OSA, 1983)	84-71539	0-88318-317-X
No. 119	Laser Techniques in the Extreme Ultraviolet (OSA, Boulder, CO, 1984)	84-72128	0-88318-318-8
No. 120	Optical Effects in Amorphous Semiconductors (Snowbird, UT, 1984)	84-72419	0-88318-319-6

No. 121	High Energy e^+e^- Interactions (Vanderbilt, 1984)	84-72632	0-88318-320-X
No. 122	The Physics of VLSI (Xerox, Palo Alto, CA, 1984)	84-72729	0-88318-321-8
No. 123	Intersections Between Particle and Nuclear Physics (Steamboat Springs, CO, 1984)	84-72790	0-88318-322-6
No. 124	Neutron-Nucleus Collisions: A Probe of Nuclear Structure (Burr Oak State Park, 1984)	84-73216	0-88318-323-4
No. 125	Capture Gamma-Ray Spectroscopy and Related Topics – 1984 (Int'l Symposium, Knoxville, TN)	84-73303	0-88318-324-2
No. 126	Solar Neutrinos and Neutrino Astronomy (Homestake, 1984)	84-63143	0-88318-325-0
No. 127	Physics of High Energy Particle Accelerators (BNL/SUNY Summer School, 1983)	85-70057	0-88318-326-9
No. 128	Nuclear Physics with Stored, Cooled Beams (McCormick's Creek State Park, IN, 1984)	85-71167	0-88318-327-7
No. 129	Radiofrequency Plasma Heating (Sixth Topical Conference) (Callaway Gardens, GA, 1985)	85-48027	0-88318-328-5
No. 130	Laser Acceleration of Particles (Malibu, CA, 1985)	85-48028	0-88318-329-3
No. 131	Workshop on Polarized ^{3}He Beams and Targets (Princeton, NJ, 1984)	85-48026	0-88318-330-7
No. 132	Hadron Spectroscopy–1985 (International Conference, Univ. of Maryland)	85-72537	0-88318-331-5
No. 133	Hadronic Probes and Nuclear Interactions (Arizona State University, 1985)	85-72638	0-88318-332-3
No. 134	The State of High Energy Physics (BNL/SUNY Summer School, 1983)	85-73170	0-88318-333-1
No. 135	Energy Sources: Conservation and Renewables (APS, Washington, DC, 1985)	85-73019	0-88318-334-X
No. 136	Atomic Theory Workshop on Relativistic and QED Effects in Heavy Atoms (Gaithersburg, MD, 1985)	85-73790	0-88318-335-8
No. 137	Polymer-Flow Interaction (La Jolla Institute, 1985)	85-73915	0-88318-336-6
No. 138	Frontiers in Electronic Materials and Processing (Houston, TX, 1985)	86-70108	0-88318-337-4
No. 139	High-Current, High-Brightness, and High-Duty Factor Ion Injectors (La Jolla Institute, 1985)	86-70245	0-88318-338-2
No. 140	Boron-Rich Solids (Albuquerque, NM, 1985)	86-70246	0-88318-339-0
No. 141	Gamma-Ray Bursts (Stanford, CA, 1984)	86-70761	0-88318-340-4
No. 142	Nuclear Structure at High Spin, Excitation, and Momentum Transfer (Indiana University, 1985)	86-70837	0-88318-341-2
No. 143	Mexican School of Particles and Fields (Oaxtepec, México, 1984)	86-81187	0-88318-342-0
No. 144	Magnetospheric Phenomena in Astrophysics (Los Alamos, NM, 1984)	86-71149	0-88318-343-9

No. 145	Polarized Beams at SSC & Polarized Antiprotons (Ann Arbor, MI & Bodega Bay, CA, 1985)	86-71343	0-88318-344-7
No. 146	Advances in Laser Science–I (Dallas, TX, 1985)	86-71536	0-88318-345-5
No. 147	Short Wavelength Coherent Radiation: Generation and Applications (Monterey, CA, 1986)	86-71674	0-88318-346-3
No. 148	Space Colonization: Technology and The Liberal Arts (Geneva, NY, 1985)	86-71675	0-88318-347-1
No. 149	Physics and Chemistry of Protective Coatings (Universal City, CA, 1985)	86-72019	0-88318-348-X
No. 150	Intersections Between Particle and Nuclear Physics (Lake Louise, Canada, 1986)	86-72018	0-88318-349-8
No. 151	Neural Networks for Computing (Snowbird, UT, 1986)	86-72481	0-88318-351-X
No. 152	Heavy Ion Inertial Fusion (Washington, DC, 1986)	86-73185	0-88318-352-8
No. 153	Physics of Particle Accelerators (SLAC Summer School, 1985) (Fermilab Summer School, 1984)	87-70103	0-88318-353-6
No. 154	Physics and Chemistry of Porous Media—II (Ridge Field, CT, 1986)	83-73640	0-88318-354-4
No. 155	The Galactic Center: Proceedings of the Symposium Honoring C. H. Townes (Berkeley, CA, 1986)	86-73186	0-88318-355-2
No. 156	Advanced Accelerator Concepts (Madison, WI, 1986)	87-70635	0-88318-358-0
No. 157	Stability of Amorphous Silicon Alloy Materials and Devices (Palo Alto, CA, 1987)	87-70990	0-88318-359-9
No. 158	Production and Neutralization of Negative Ions and Beams (Brookhaven, NY, 1986)	87-71695	0-88318-358-7
No. 159	Applications of Radio-Frequency Power to Plasma: Seventh Topical Conference (Kissimmee, FL, 1987)	87-71812	0-88318-359-5
No. 160	Advances in Laser Science–II (Seattle, WA, 1986)	87-71962	0-88318-360-9
No. 161	Electron Scattering in Nuclear and Particle Science: In Commemoration of the 35th Anniversary of the Lyman-Hanson-Scott Experiment (Urbana, IL, 1986)	87-72403	0-88318-361-7
No. 162	Few-Body Systems and Multiparticle Dynamics (Crystal City, VA, 1987)	87-72594	0-88318-362-5
No. 163	Pion–Nucleus Physics: Future Directions and New Facilities at LAMPF (Los Alamos, NM, 1987)	87-72961	0-88318-363-3
No. 164	Nuclei Far from Stability: Fifth International Conference (Rosseau Lake, ON, 1987)	87-73214	0-88318-364-1
No. 165	Thin Film Processing and Characterization of High-Temperature Superconductors (Anaheim, CA, 1987)	87-73420	0-88318-365-X

No. 166	Photovoltaic Safety (Denver, CO, 1988)	88-42854	0-88318-366-8
No. 167	Deposition and Growth: Limits for Microelectronics (Anaheim, CA, 1987)	88-71432	0-88318-367-6
No. 168	Atomic Processes in Plasmas (Santa Fe, NM, 1987)	88-71273	0-88318-368-4
No. 169	Modern Physics in America: A Michelson-Morley Centennial Symposium (Cleveland, OH, 1987)	88-71348	0-88318-369-2
No. 170	Nuclear Spectroscopy of Astrophysical Sources (Washington, DC, 1987)	88-71625	0-88318-370-6
No. 171	Vacuum Design of Advanced and Compact Synchrotron Light Sources (Upton, NY, 1988)	88-71824	0-88318-371-4
No. 172	Advances in Laser Science–III: Proceedings of the International Laser Science Conference (Atlantic City, NJ, 1987)	88-71879	0-88318-372-2
No. 173	Cooperative Networks in Physics Education (Oaxtepec, Mexico, 1987)	88-72091	0-88318-373-0
No. 174	Radio Wave Scattering in the Interstellar Medium (San Diego, CA, 1988)	88-72092	0-88318-374-9
No. 175	Non-neutral Plasma Physics (Washington, DC, 1988)	88-72275	0-88318-375-7
No. 176	Intersections Between Particle and Nuclear Physics (Third International Conference) (Rockport, ME, 1988)	88-62535	0-88318-376-5
No. 177	Linear Accelerator and Beam Optics Codes (La Jolla, CA, 1988)	88-46074	0-88318-377-3
No. 178	Nuclear Arms Technologies in the 1990s (Washington, DC, 1988)	88-83262	0-88318-378-1
No. 179	The Michelson Era in American Science: 1870–1930 (Cleveland, OH, 1987)	88-83369	0-88318-379-X
No. 180	Frontiers in Science: International Symposium (Urbana, IL, 1987)	88-83526	0-88318-380-3
No. 181	Muon-Catalyzed Fusion (Sanibel Island, FL, 1988)	88-83636	0-88318-381-1
No. 182	High T_c Superconducting Thin Films, Devices, and Application (Atlanta, GA, 1988)	88-03947	0-88318-382-X
No. 183	Cosmic Abundances of Matter (Minneapolis, MN, 1988)	89-80147	0-88318-383-8
No. 184	Physics of Particle Accelerators (Ithaca, NY, 1988)	89-83575	0-88318-384-6
No. 185	Glueballs, Hybrids, and Exotic Hadrons (Upton, NY, 1988)	89-83513	0-88318-385-4
No. 186	High-Energy Radiation Background in Space (Sanibel Island, FL, 1987)	89-83833	0-88318-386-2
No. 187	High-Energy Spin Physics (Minneapolis, MN, 1988)	89-83948	0-88318-387-0
No. 188	International Symposium on Electron Beam Ion Sources and their Applications (Upton, NY, 1988)	89-84343	0-88318-388-9

No. 189	Relativistic, Quantum Electrodynamic, and Weak Interaction Effects in Atoms (Santa Barbara, CA, 1988)	89-84431	0-88318-389-7
No. 190	Radio-frequency Power in Plasmas (Irvine, CA, 1989)	89-45805	0-88318-397-8
No. 191	Advances in Laser Science–IV (Atlanta, GA, 1988)	89-85595	0-88318-391-9
No. 192	Vacuum Mechatronics (First International Workshop) (Santa Barbara, CA, 1989)	89-45905	0-88318-394-3
No. 193	Advanced Accelerator Concepts (Lake Arrowhead, CA, 1989)	89-45914	0-88318-393-5
No. 194	Quantum Fluids and Solids—1989 (Gainesville, FL, 1989)	89-81079	0-88318-395-1
No. 195	Dense Z-Pinches (Laguna Beach, CA, 1989)	89-46212	0-88318-396-X
No. 196	Heavy Quark Physics (Ithaca, NY, 1989)	89-81583	0-88318-644-6
No. 197	Drops and Bubbles (Monterey, CA, 1988)	89-46360	0-88318-392-7
No. 198	Astrophysics in Antarctica (Newark, DE, 1989)	89-46421	0-88318-398-6
No. 199	Surface Conditioning of Vacuum Systems (Los Angeles, CA, 1989)	89-82542	0-88318-756-6
No. 200	High T_c Superconducting Thin Films: Processing, Characterization, and Applications (Boston, MA, 1989)	90-80006	0-88318-759-0
No. 201	QED Stucture Functions (Ann Arbor, MI, 1989)	90-80229	0-88318-671-3
No. 202	NASA Workshop on Physics From a Lunar Base (Stanford, CA, 1989)	90-55073	0-88318-646-2
No. 203	Particle Astrophysics: The NASA Cosmic Ray Program for the 1990s and Beyond (Greenbelt, MD, 1989)	90-55077	0-88318-763-9
No. 204	Aspects of Electron–Molecule Scattering and Photoionization (New Haven, CT, 1989)	90-55175	0-88318-764-7
No. 205	The Physics of Electronic and Atomic Collisions (XVI International Conference) (New York, NY, 1989)	90-53183	0-88318-390-0
No. 206	Atomic Processes in Plasmas (Gaithersburg, MD, 1989)	90-55265	0-88318-769-8
No. 207	Astrophysics from the Moon (Annapolis, MD, 1990)	90-55582	0-88318-770-1
No. 208	Current Topics in Shock Waves (Bethlehem, PA, 1989)	90-55617	0-88318-776-0
No. 209	Computing for High Luminosity and High Intensity Facilities (Santa Fe, NM, 1990)	90-55634	0-88318-786-8
No. 210	Production and Neutralization of Negative Ions and Beams (Brookhaven, NY, 1990)	90-55316	0-88318-786-8
No. 211	High-Energy Astrophysics in the 21st Century (Taos, NM, 1989)	90-55644	0-88318-803-1

No. 212	Accelerator Instrumentation (Brookhaven, NY, 1989)	90-55838	0-88318-645-4
No. 213	Frontiers in Condensed Matter Theory (New York, NY, 1989)	90-6421	0-88318-771-X 0-88318-772-8 (pbk.)
No. 214	Beam Dynamics Issues of High-Luminosity Asymmetric Collider Rings (Berkeley, CA, 1990)	90-55857	0-88318-767-1
No. 215	X-Ray and Inner-Shell Processes (Knoxville, TN, 1990)	90-84700	0-88318-790-6
No. 216	Spectral Line Shapes, Vol. 6 (Austin, TX, 1990)	90-06278	0-88318-791-4
No. 217	Space Nuclear Power Systems (Albuquerque, NM, 1991)	90-56220	0-88318-838-4
No. 218	Positron Beams for Solids and Surfaces (London, Canada, 1990)	90-56407	0-88318-842-2
No. 219	Superconductivity and Its Applications (Buffalo, NY, 1990)	91-55020	0-88318-835-X
No. 220	High Energy Gamma-Ray Astronomy (Ann Arbor, MI, 1990)	91-70876	0-88318-812-0
No. 221	Particle Production Near Threshold (Nashville, IN, 1990)	91-55134	0-88318-829-5
No. 222	After the First Three Minutes (College Park, MD, 1990)	91-55214	0-88318-828-7
No. 223	Polarized Collider Workshop (University Park, PA, 1990)	91-71303	0-88318-826-0
No. 224	LAMPF Workshop on (π, K) Physics (Los Alamos, NM, 1990)	91-71304	0-88318-825-2
No. 225	Half Collision Resonance Phenomena in Molecules (Caracus, Venezuela, 1990)	91-55210	0-88318-840-6
No. 226	The Living Cell in Four Dimensions (Gif sur Yvette, France, 1990)	91-55209	0-88318-794-9
No. 227	Advanced Processing and Characterization Technologies (Clearwater, FL, 1991)	91-55194	0-88318-910-0
No. 228	Anomalous Nuclear Effects in Deuterium/Solid Systems (Provo, UT, 1990)	91-55245	0-88318-833-3
No. 229	Accelerator Instrumentation (Batavia, IL, 1990)	91-55347	0-88318-832-1
No. 230	Nonlinear Dynamics and Particle Acceleration (Tsukuba, Japan, 1990)	91-55348	0-88318-824-4
No. 231	Boron-Rich Solids (Albuquerque, NM, 1990)	91-53024	0-88318-793-4
No. 232	Gamma-Ray Line Astrophysics (Paris–Saclay, France, 1990)	91-55492	0-88318-875-9
No. 233	Atomic Physics 12 (Ann Arbor, MI, 1990)	91-55595	0-88318-811-2
No. 234	Amorphous Silicon Materials and Solar Cells (Denver, CO, 1991)	91-55575	0-88318-831-7

No.	Title		
No. 235	Physics and Chemistry of MCT and Novel IR Detector Materials (San Francisco, CA, 1990)	91-55493	0-88318-931-3
No. 236	Vacuum Design of Synchrotron Light Sources (Argonne, IL, 1990)	91-55527	0-88318-873-2
No. 237	Kent M. Terwilliger Memorial Symposium (Ann Arbor, MI, 1989)	91-55576	0-88318-788-4
No. 238	Capture Gamma-Ray Spectroscopy (Pacific Grove, CA, 1990)	91-57923	0-88318-830-9
No. 239	Advances in Biomolecular Simulations (Obernai, France, 1991)	91-58106	0-88318-940-2
No. 240	Joint Soviet-American Workshop on the Physics of Semiconductor Lasers (Leningrad, USSR, 1991)	91-58537	0-88318-936-4
No. 241	Scanned Probe Microscopy (Santa Barbara, CA, 1991)	91-76758	0-88318-816-3
No. 242	Strong, Weak, and Electromagnetic Interactions in Nuclei, Atoms, and Astrophysics: A Workshop in Honor of Stewart D. Bloom's Retirement (Livermore, CA, 1991)	91-76876	0-88318-943-7
No. 243	Intersections Between Particle and Nuclear Physics (Tucson, AZ, 1991)	91-77580	0-88318-950-X
No. 244	Radio Frequency Power in Plasmas (Charleston, SC, 1991)	91-77853	0-88318-937-2
No. 245	Basic Space Science (Bangalore, India, 1991)	91-78379	0-88318-951-8
No. 246	Space Nuclear Power Systems (Albuquerque, NM, 1992)	91-58793	1-56396-027-3 1-56396-026-5 (pbk.)
No. 247	Global Warming: Physics and Facts (Washington, DC, 1991)	91-78423	0-88318-932-1
No. 248	Computer-Aided Statistical Physics (Taipei, Taiwan, 1991)	91-78378	0-88318-942-9
No. 249	The Physics of Particle Accelerators (Upton, NY, 1989, 1990)	XX-XXXXX	0-88318-789-2
No. 250	Towards a Unified Picture of Nuclear Dynamics (Nikko, Japan, 1991)	92-70143	0-88318-951-8
No. 251	Superconductivity and its Applications (Buffalo, NY, 1991)	92-52726	1-56396-016-8
No. 252	Accelerator Instrumentation (Newport News, VA, 1991)	92-70356	0-88318-934-8
No. 253	High-Brightness Beams for Advanced Accelerator Applications (College Park, MD, 1991)	92-52705	0-88318-947-X
No. 254	Testing the AGN Paradigm (College Park, MD, 1991)	92-52780	1-56396-009-5
No. 255	Advanced Beam Dynamics Workshop on Effects of Errors in Accelerators, Their Diagnosis and Corrections (Corpus Christi, TX, 1991)	XX-XXXXX	1-56396-006-0